HORSEMANSHIP AND HORSEMASTERSHIP
Volume 1

XENOPHON PRESS

HORSEMANSHIP AND HORSEMASTERSHIP
Volume 1

Horsemanship Part I:
The Education of The Rider

&

Horsemastership Part II:
The Education of the Horse

Copyright © 1988 by the U.S. Horse Cavalry Association.

All rights reserved. No part of this publication my be reproduced, stored in a retrieval system, or transmitted in any form or by any means, electronic, mechanical, photocopying, digital, recording or otherwise with the prior written permission of the publisher.

Published by Xenophon Press LLC
XenophonPress@gmail.com
PO Box 137 Nassawadox Va 23413
757-442-6769.

Library of Congress Catalog Card Number 88-051235

Xenophon Press Edition
Copyright 2021

ISBN: 978-1948717328

PUBLISHER'S PREFACE

These two volumes, in one book, are veritable classics of the American equestrian tradition. Originally written as manuals used to train recruits and remounts (horses) for the Cavalry, they contain timeless information and methods that can be applied by horsemen and horsewomen of all disciplines today. The truths contained in these pages are universal. These texts were written to describe a system by which ordinary lay people (recruits) could be taught to ride in a pragmatic, easy-to-follow method. Further, the method for teaching *Horsemanship: Part I* was devised so that it did not rely on the rider's inherent talent for riding (or lack thereof). The beauty of the method, is that it is democratic and pragmatic in its approach. Similarly, the training of the horse outlined in *Horsemastership: Part II*, is applicable to grade horses irrespective of their innate ride-ability or train-ability. The Cavalry came by horses by a variety of means and the stock was of varied talent. The methods outlined in Part II when applied, and followed, will work with most any horse. What was true 80-100 years ago is still relevant today; equine nature has not changed in the past 100 years.

I want to thank the U.S. Cavalry Association for entrusting Xenophon Press to bring this book back into print, making it available worldwide through our distribution network. More specifically, I want to thank Karen Tempero for her excellent editorial suggestions which have helped me, as the editor, correct errors, and bring some of the older syntax up to date. Xenophon Press is proud to add *Horsemanship & Horsemastership* to our ever-growing Xenophon Press Library. We have enlarged the font size, improved the images where possible, and produced a hardcover collector's edition. Xenophon Press has a multi-year goal of preserving the classics of equestrian literature. This includes the keystones of American riding texts such as this work, and the texts by one of its contributors, Harry D. Chamberlin.

We hope you enjoy, and more importantly, are able to make use of the material represented here.

Richard F. Williams
Editor-in-Chief
Xenophon Press

FOREWORD

A classic in its field and once an official textbook of the U.S. Cavalry School, located in Fort Riley, Kansas, this manual influenced the professional development of horse Cavalrymen during the last years of the mounted services in U.S. Army. Long out of print, its reappearance will be welcomed by all who share an interest in horse lore, and especially by Cavalry veterans who refined their own equestrian skills in accordance with its principles.

Those principles are as valid today as when they were being taught to our military horsemen between the two world wars. Consequently, this manual, together with two companion manuals of the Cavalry School revived to form a series, can be of great value to anyone interested in riding, schooling and managing horses.

The U.S. Cavalry Association, as conservator of the history and custodian of the traditions of our mounted services, is proud to make these manuals available once again in partnership with Xenophon Press.

Membership in the U.S. Cavalry Association is open to all who are interested in the preservation of the history and heritage of the Horse Cavalry and Horse Artillery of the U.S. Army. The Association collects Cavalry artifacts, preserves historical records, sponsors historical publications, disseminates information about the Cavalry, produces visual materials on Cavalry subjects, and conducts meetings and reunions.

For membership contact:
info@uscavalryassociation.org
U.S. Cavalry Association
3220 N. Jesse Reno Street
El Reno, OK 73036
www.uscavalryassociation.org
405-422-6330

INTRODUCTION TO THE XENOPHON PRESS EDITION

This 1942 edition of *Horsemanship and Horsemastership, Volume I, Part One—Education of the Rider; Part Two— Education of the Horse* introduces you to the principles of riding and training developed by the US Cavalry Board, Brigadier General Harry D. Chamberlin, and other cavalrymen many of whose names are lost to history. As James Wofford makes clear in his Forward to the biography *General Chamberlin: America's Equestrian Genius* [Xenophon Press 2020], these principles were taught not only to him but also to Gordon Wright, George H. Morris, William Steinkraus, Frank Chapot, Michael Plumb, Michael Page, Kevin Freeman and a host of others. This manual distills learning acquired at the world's finest schools of horsemanship in France, Germany, Italy, Poland, Sweden, Austria, and the United States to name only a few. The manual integrates this learning into a uniquely American system of riding and training.

To apply several points that General John Tupper Cole wrote about Harry Chamberlin's method of riding, one accurately might say: In the matter-of-fact and workmanlike manner of a military manual, *"H and H*[1]*"* especially integrates the dressage of Saumur, the modified jumping concepts of Tor di Quinto, the accuracy of Hanover, and the long distance riding concepts of Ft. Riley. And this remains significant since, as so often happened in Europe, national pride and the "not invented here" mentality often isolated the horseman of each country from learning the insights of their neighbors.

Not so for the American cavalry officers who studied at those schools. They studied all, evaluated all, and copied from all—and when they returned to America, they refined and modified what they had learned to shape the integrated system you see in *"H and H"* to further the goals of the United States Cavalry: to train horse

---

1   Horsemanship and Horsemastership

and rider to travel long distances and to surmount all obstacles; and most of all, to enable riders to get the most *from* the horse while taking the least *out* of the horse.

As you read, you will see immediately the influence of Brigadier General Harry D. Chamberlin and his two books *Riding and Schooling Horses* and *Training Hunters, Jumpers, and Hacks* [Xenophon Press 2020]. Chamberlin remained the leading light of the US Cavalry's equestrian thinking from 1922 to 1942. While Commandant of Ft. Riley's Cavalry Replacement Training Center and as the US Cavalry's greatest horseman, Chamberlin appears in this edition of *Horsemanship and Horsemastership*. Chamberlin demonstrates in photos and describes in words the military seat also known in the Army as the Chamberlin military seat.

You also will see the influence of the masters of French *dressage* such as General Alexis L'Hotte, the Count D'Aure, and François Baucher as interpreted by Faverot de Kerbrech, and James Fillis—but modified to suit the demands of the US Cavalry's mode of riding. Unlike the Italian Caprillists[2] who rejected French principles, General Chamberlin, and as a result the entire US Cavalry's officer corps, took French dressage seriously given the American need for horses to be "handy" and "nimble" for maneuvers. At the same time, to preserve the natural balance of the horse for cross country and distance riding, Chamberlin applied French principles selectively always keeping in mind that *dressage* served as a means to an end, not as an end itself. Nonetheless, you will see Caprilli's influence, especially to educate the rider, but filtered and modified by the genius of Harry Chamberlin. For the most part, however, Chamberlin rejected pure Caprillist concepts to educate the horse in favor of French doctrine—and that preference clearly shows in this manual.

During World War II, this 1942 edition of *"H and H"* (especially *Part One—Education of the Rider*) served as the textbook for recruits at Ft. Riley's Cavalry Replacement Training Center (the CRTC). As students at the CRTC, riding instructor Gordon Wright, future Olympic gold medalist William Steinkraus, and others who molded American equestrianism studied directly from the pages of this manual. As William Steinkraus wrote: "At the CRTC, they taught me how to ride all over again." As Gordon Wright maintained: this manual presents a step by step approach and offers an "inspiration" to all who aspire to excellent riding. And through Gordon Wright this manual's principles passed to the elite of America's equestrian world.

Actually, the entire series *Horsemanship and Horsemastership, Volume I: Part One—Education of the Rider; Part Two— Education of the Horse; Volume II: Animal Management; and Volume III: Horse Shows, Race Meetings, Hunter Trials,*

---

2   [followers of Federico Caprilli, the Italian cavalry officer who revolutionized the way horse riders jump fences. - Editor's note.]

Mounted Sports, Exhibition Rides presents the efforts of the US Cavalry Board, of Brigadier General Chamberlin. and of the other American cavalrymen who traveled the world in a quest for equestrian knowledge. This series served as the US Cavalry's "equestrian bible." It educated and inspired the founders of the civilian United States Equestrian Team. It educated and inspired so many others who exerted so much influence on American equestrian sport. Chamberlin's two books combined with this manual provide the foundation for America's success in Eventing and Stadium Jumping in Olympic, Pan American and other international competition. For eventers, *Grand Prix* riders, endurance and long distance riders, mounted police units, military units, re-enactment groups and others, Xenophon Press' publication of this 1942 version of *Horsemanship and Horsemastership, Volume I* and of the two Chamberlin books provides not only a bit of history but also great resources for novices, experienced horsemen, and Olympians alike to hone their skills.

Warren C. Matha,
February 2021

**The Cavalry School
Fort Riley, Kansas
1945**

This publication, prepared by the Academic Division's Department of Horsemanship, is a text for use in residence courses at The Cavalry School. Published in its present form in 1942 under the direction of Colonel Dorsey R. Rodney, Cavalry, then Assistant Commandant, *Horsemanship and Horsemastership* has increased steadily in popularity and influence. It has proved valuable and interesting to horsemen, both military and civilian, throughout the country.

>John B. Thompson,
>*Colonel, Cavalry,*
>*Assistant Commandant.*

CONTENTS: PART I

Publisher's Preface v
Foreword vi
Introduction to the Xenophon Press Edition vii
Message from John B. Thompson 1945.. xi

Horsemanship Part I : The Education of The Rider 1
I: General Considerations 1 3
II: Knowledge and Utilization of the Mental and
      Moral Faculties of the Horse 2-7 4
III: Articles of Equipment and Their Adjustment 8-12................ 9
IV: The Position of the Rider 13-16... 18
V: Suppling Exercises 17-19... 36
VI: The Application and Action of Aids 20-25........................... 44
VII: Management and Control of the Horse 26-49...................... 67
VIII: Riding Hall Movements and Customs 50-51 113
IX: Jumping 62-69 123
X: Cross Country Riding 60-67... 140
XI: Suggestions for Instruction 68-69 144

HORSEMANSHIP PART I: THE EDUCATION OF THE RIDER

CHAPTER I
GENERAL CONSIDERATIONS

1. The purpose of elementary equitation is to give the rider a firm and correct seat and to enable him to acquire sufficient authority to control his horse under the normal conditions in which he is used, such as pleasure riding, work with arms, hunting, etc. The use of the horse under these circumstances demands on the part of the rider the knowledge and application of a certain number of methods which enables him to indicate his will and to cause the horse to submit thereto. For control a study must be made of the mental faculties of the horse as well as the methods by which these faculties may be utilized through the use of the aids; that is, by the application of the reins, legs and weight of the rider.

CHAPTER II

KNOWLEDGE AND UTILIZATION OF THE MENTAL AND MORAL FACULTIES OF THE HORSE

    2. Memory 4
    3. Confidence and Fear 4
    4. Comparison of the Sensations 5
    5. Willingness 5
    6. Rewards 6
    7. Punishment 6

2. MEMORY.

The horse, to a great degree, owes his aptitude for training to memory. His training has, in fact, been accomplished because he remembers or recognizes the indications given him, the manner in which he responded, and finally the rewards or the punishments which followed. This fact should be taken into account not only during training, but every time the trained horse is used.

    The horse is one of the most eccentric of animals, and very easily acquires a good or bad habit, therefore, the rider must be very careful to prevent a fault being committed several times uncorrected, as it will become a habit which, although easy to correct at the start, often is eliminated only with great difficulty if allowed to continue for any length of time.

    Similarly the rider should carefully avoid employing incorrect, though momentarily easier, methods with the intention of later substituting others, as this causes much loss of time and the subsequent correction is often difficult.

3. CONFIDENCE AND FEAR.

The horse's intelligence is too rudimentary for him to understand or calmly accept the out-of-door phenomena which affect him. Strange objects frequently frighten him and this fear is increased when he is improperly handled; therefore, this tendency of fear should not be developed by brusque or violent methods, but should be lessened by utilizing the means with which the horse is calmed. Calmness and confidence are gained only when the rider, by sympathetic yet firm methods, forces the horse to overcome his natural apprehensiveness.

If the horse has the tendency to fear everything strange that he encounters, the rider must gradually overcome this fear in order to assert his mastery. Herein is the role of punishment, which should be given only at the proper time and to the exact degree required, and which should be inflicted only, and at the instant, when the horse is voluntarily and knowingly disobedient. He is incapable of associating reward or punishment with an act of obedience or disobedience, unless such reward or punishment *immediately* follows the act.

It can thus be seen that the rider must be perfectly calm and master of himself in order to acquire the patience by which he gains the horse's confidence.

4. COMPARISONS OF THE SENSATIONS.

By the comparison of sensations, and assisted by his memory, the horse distinguishes the phenomena taking place around him. Thus, when given oats shortly after having heard the grain bin opened, the horse associates these two sensations, viz.—the sound of the opening cover and the pleasure of receiving oats.

This fact is proved by his whinnying when he hears the above sound. This association of ideas is taken advantage of in many ways and, by means of it, significance is given to rewards and punishments.

The rider must therefore remember that his horse is always disposed to respond to the same demands in the same way. Consequently, in training, as well as in the practice of equitation, care must be taken that the horse executes the movement demanded as soon as practicable, or at least progresses toward the goal set.

5. WILLINGNESS.

All horses have this faculty in varying amounts. Some submit to the rider's control with little trouble and hesitation, while others, on the contrary, offer great resistance. Under certain circumstances all horses display stubbornness. The rider must always expect to encounter this stubbornness, which he must seek to avoid by the employment of the proper aids; by demanding of the horse only what he is able to do, taking into account his degree of training and his physical attributes. Moreover, the rider must judiciously employ rewards and punishments to assure his supremacy over the will of the horse.

6. REWARD.

The horse is susceptible to rewards and understands their meaning. They stimulate his desire to please, encourage and reassure him when frightened by an unknown demand, and restore his confidence and submission. By these means the rider causes the horse to repeat movements already obtained, and, together with the voice, they may obtain the best results. Rewards should not, however, be given the horse promiscuously. Riders are often seen caressing horses which are completely insubordinate. This is a bad error. If the horse, upset through fear or ignorance, becomes excited and does not allow himself to be controlled by the aids to the degree demanded, he should be calmed by use of the voice and by stroking. But if he knows what is wanted of him, and for no reason whatsoever resists, it is an error to caress him. Such action will encourage him to resist and cause him to doubt the rider's firmness, and will necessitate sharper and more repeated corrections, and lessen the effects of rewards. Stroking the horse calms him and is a means of persuasion. It should only be employed with an excitable horse, or after the horse has yielded to some demand; never during a disobedience.

Stroking the horse is a reward which may be given often and easily, but it is not the most effective. Greatest results may be obtained through working on the horse's greediness. Satisfying his desire produces the greatest contentment and may overcome a horse's evil disposition and cause him to submit to extreme demands.

Whatever the nature of the rewards used, the rider should be free with them.

7. PUNISHMENT.

If the horse merits reward after a good performance, he also deserves punishment when he disobeys, but such action should be taken properly and at the correct time. Corrections should follow the fault immediately, in order that the horse may understand the reason for the pain he suffers. In this way only is the correction effective, for otherwise it would be misunderstood by the horse, and would be considered as an unjust and uncalled for attack.

Punishment should not only be given at the proper time but also with justice. Should the horse disobey through ignorance, fear or defects in his conformation, he should not receive punishment. This would only result in making him stubborn and disheartened, because he could not understand the reason for the punishment. If, on the contrary, the disobedience is intentional, the rider's authority must be asserted. The rider must at all costs be the master, and should not hesitate, upon proper occasion, to combat his horse. In always showing himself as the stronger, the rider takes from the horse the idea of resistance and habituates him to yielding to his demands, since the horse will recognize in him a will and means of action against which he cannot successfully cope.

Punishment should be administered without anger, as in such a case it is rarely controlled. The rider should remain calm in order to give only the needed amount of correction.

In this way he obtains greater obedience, whereas in exceeding this limit he provokes the horse's resentment and leaves with him the memory of an injustice.

As soon as the correction has produced the result sought and the horse has yielded, the rider should caress his horse in order that the latter may see that he has everything to lose by disobeying, and everything to gain by submitting. Rewards, moreover, lessen the horse's irritation, calm him, and allow the work to be continued under good conditions.

The two best means of correction are the spur and the whip, employed together or separately. Under their action the horse is obliged to move forward. While they are being employed the horse must be given a certain amount of liberty, and then picked up at the proper time. "Rein back while spurring forward" is an ironically true expression of the rider who reins back his horse while correcting him. Such action provokes resistance by the horse in place, which is the worst he can offer, and results in stubbornness.

There are occasions where the rider may dismounted give the necessary correction as for example, with horses that fall over backward, or where the rider is insufficiently experienced to punish his horse while mounted. If punishment can be inflicted immediately after the fault, it is preferable to dismount to give this punishment, rather than to remain mounted and allow the horse to go unpunished. It goes without saying that, by means of his progress, the rider tends to acquire sufficient means of support to be stronger than his horse when he is mounted.

The fact must be borne in mind that punishments are very rarely necessary. Most of the faults committed by the horse are due to his ignorance and lack of training, or to the insufficiency of the means employed by the rider. In either case, severity becomes an injustice and causes such harmful results that it is better not to punish at all than to punish wrongly.

CHAPTER III

ARTICLES OF EQUIPMENT AND THEIR ADJUSTMENT

    8. General 9
    9. Saddles 10
        a. Types 10
        b. Adjustment... 10
    10. Bridles 11
    11. Bits 11
        a. The snaffle-types and adjustment 11
        b. The curb—description and adjustment................... 12
        c. Pelham and other type bits 14
        d. Remedy for horses that put the tongue over the bit.. 14
    12. Miscellaneous equipment 15
        a. The noseband... 15
        b. Martingales 15
        c. Breast plates 16
        d. The cavesson 16
        e. The longe 17
        f. The longeing whip... 17
        g. The hackamore 17

8. GENERAL.

    a. It is beyond the scope of this manual to attempt to describe, or even to mention, all of the many articles of horse equipment. Accordingly, this section deals only with the most generally used types.

    b. The question of the proper equipment to be used depends primarily upon the characteristics of the particular horse and the purpose for which he is to be used. As a general rule, only that equipment should be used which is essential to his proper control in the work contemplated. For example, the use of the single snaffle affords adequate control in riding a steeple chase or in showing a horse over jumps. On the other hand, the use of the bit and bridoon, equipped with nose-band and standing martingale, may be essential to the proper control of the same horse on the polo field.

9. SADDLES.

a. Types.

The more commonly used saddles include the training, racing, jumping, park, polo, stock, military field, and military McClellan types. The name in each case indicates the special purpose of the type.

b. Adjustment.

(1) The saddle should be placed on the horse's back so that the front end of the side-bars are two or three inches in rear of the upper rear edges of the shoulder blades. On some horses having high thin withers, it is necessary to use a blanket, saddle pad, or pommel pads, to prevent the pommel from resting on the withers. The saddle blanket or saddle pad should be placed well forward on the horse's neck and then slid back into position so as to smooth down the hair. When the saddle is placed upon it, three or four inches of the blanket or pad should show in front of the saddle. *The blanket or pad should also extend well in rear of the saddle.* Before the saddle is girthed, the blanket or pad should be raised well up into the pommel arch to prevent injury to the withers. Pommel pads are small knitted pads used to keep the saddle off the withers. They are placed in the hollow on each side of the withers and held in place by the saddle. Most saddles have three girth straps and most girths have two buckles. With narrow chested, large bellied horses, the girth should be buckled on the rear straps. This will ordinarily prevent galling. It may, however, be necessary in some cases to cover the girth with sheepskin. Girths are ordinarily made of leather, cord, or cotton or woolen webbing. The leather girth should be used with the folded edge to the front, and should be kept soft by unfolding and treating with oil. The girth, when first adjusted, should admit a finger between it and the horse's belly. After the horse has had a few minutes exercise, the girth should always be examined and tightened if necessary.

(2) The rider may, before mounting, approximate the correct length of stirrup straps by adjusting them so that the length of the strap, including the stirrup, is about one inch less than the length of the arm with the

fingers extended. After mounting, the rider should adjust the stirrup straps accurately.

10. BRIDLES.

Bridles may be classified as single and double. A single bridle is equipped with one bit; whereas a double bridle is ordinarily equipped with two snaffles or a bit and bridoon, two head stalls, and two pairs of reins.

All bridles should be fitted so that the crown piece neither slides back on the horse's neck, nor pulls up against his ears. The throat-latch should be buckled loosely enough to permit the hand, held in a vertical plane, to be passed between it and the horse's throat.

11. BITS.

There are more varieties of bits than of any other article of horse equipment. The most common types are the snaffle, the curb, and the Pelham. In general, the manner in which the bit is designed to act on the horse's mouth determines the type. Bits of various degrees of severity are to be found in all types.

a. The snaffle. Types. The ordinary snaffle is a bit composed of two slightly curved and conical pieces of metal joined at their small ends. These pieces are so joined that they are movable with respect to each other. In the larger end of each part is a hole through which is passed a metallic ring of variable size, but large enough for the reins and cheek straps to be attached thereto. Different methods are taken to prevent this ring from entering the horse's mouth. The principal ones are the following:

Circular guards of rubber or leather at each end.

Metal branches tangential to and forming part of the ring. These extend beyond the two sides and act across and against the animal's lips if the ring tends to enter the mouth.

Metal branch similar to the above extending beyond on one side only and having a loop at its end to serve as the attachment of the cheek strap. This snaffle bit is called the "Baucher Snaffle."

With horses having very sensitive mouths, thick snaffles, straight bar snaffles, or those covered with leather or rubber may be used.

The snaffle is a very mild bit acting mainly on the lips and slightly on the bars. The rider should use this bit in the beginning as its mildness renders the faults of the hands less harmful to the horse's mouth.

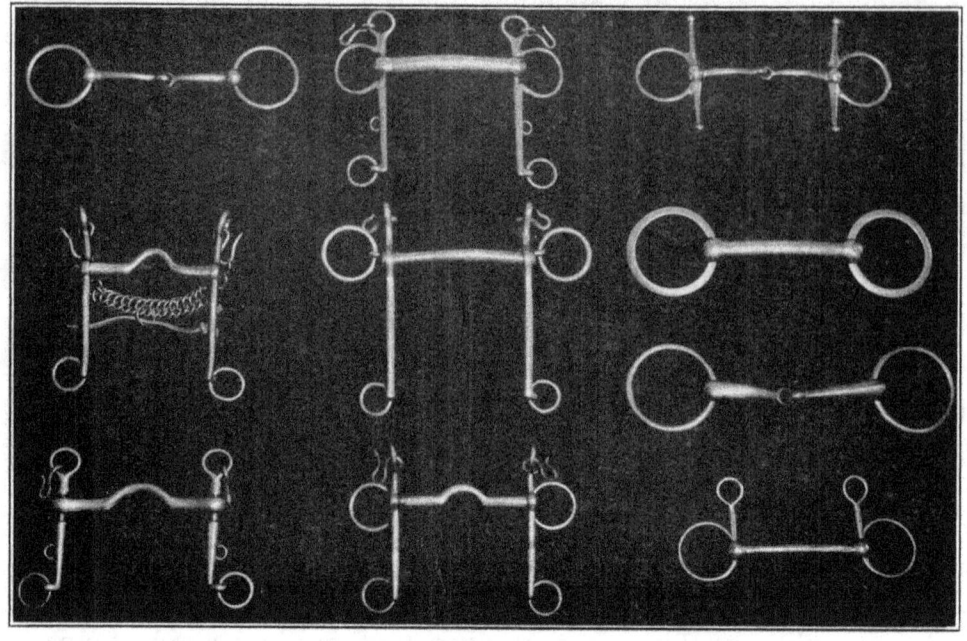

BITS

Snaffle.
Curb with Chain
  and Lip Thong.
Curb, Tom Thumb
  Type.

Pelham, Half Moon
  Sliding Bar.
Pelham, Half Moon.
Pelham, Straight Bar
  with Port.

Snaffle, with Cheek Bars.
Snaffle, Straight Bar.
Snaffle, Hunting.
Snaffle, Baucher with
  Straight Bar.

Fig. 1.

(2) *Adjustment.*

The single snaffle should be adjusted so that it fits snugly in the corners of the horse's mouth. A shorter adjustment causes several wrinkles at the corners of the mouth, whereas a longer adjustment allows the bit to strike the tushes. Either may cause pain and fretfulness. The double snaffle is used in preparation for the bit and bridoon. When used the upper snaffle is attached to the inner or snaffle cheek pieces and should be adjusted like the single snaffle. The lower snaffle, corresponding to the curb, is attached to the outer or curb cheek pieces and should rest on the bars of the mouth immediately below the upper snaffle. A double snaffle consisting of a branch snaffle and a "Baucher snaffle" is the most effective combination.

b. The Curb Bit.

(1) *Description.* The curb bit is composed of two branches joined by a mouthpiece. The cheek straps are fastened to a small opening on one end of each branch. The other end of each branch has fastened to it a movable ring to which the curb rein is attached. These two branches are joined by

the mouthpiece, the length of which varies according to the size of the mouth for which it is intended. Each end of this mouthpiece is called a cannon and rests on the bars of the horse's mouth. The cannons are cylindrical and may consist of one straight piece, or be separated by a rounded shoulder called the port, in which the thick part of the tongue may rest. This shoulder allows the cannons to act with greater force on the bars. From this we deduce that the more liberty of tongue allowed, the more severe becomes the bit. The mildest bits, called straight bits, are those which allow no liberty to the tongue. The curb chain attached to the upper part of the branches by hooks, completes this bit.

The curb bit is designed to act on the bars of the horse's mouth, rather than on the lips and tongue, and by leverage rather than by direct pressure. This action is dependent upon the curb chain (or curb strap), which passes under the horse's lower jaw. The curb is essentially a severe bit, which, for this reason, is seldom used alone.

The severity of the bit is increased by:
Increasing the ratio between the length of the lower and upper branches.
Increasing the tightness of the curb chain.
Increasing the liberty accorded the tongue.
Decreasing the thickness of the cannons.

There are many varieties and types of curb bits varying in their degrees of severity. One of the most useful types is the sliding bar bit in which the cannons slide on the branches with about a half an inch of play. This is a milder bit than the ordinary curb and is useful in getting horses to relax their jaws.

Bit and bridoon is a term applied to the curb and snaffle used in combination, in which bit refers to the curb and bridoon to the snaffle. Not only is it the standard type for the military service, but it also is the one most frequently used for park riding, cross country riding, schooling and training horses, and playing polo.

(2) *Adjustment.*

The bit and bridoon should be adjusted as follows: The Snaffle in the same manner as the single snaffle, and The Curb so that it rests on the bars of the mouth just below the snaffle. It is hard to prescribe exactly how far below the snaffle the curb should rest; it should be at least one-half inch above the tushes, or two inches above the corner incisors, when there are no tushes. Care should be taken to insure that this clearance of the tushes affects the uppers as well as the lowers. The lower it is adjusted, the greater is its severity. The mouthpiece should be of such breadth that the branches will lightly touch the sides of the lips. The chain should be twisted until flat and should then be hooked at such a length that the branches of the bit will form an angle of forty-five degrees with the lower jaw when the reins are lightly stretched. The chain should pass below, not above, the snaffle. If the curb chain is too loose, the bit "falls through" and loses its curb effect and acts as the snaffle on the corners of the horse's mouth. If the chain is too tight the horse be comes over-flexed and fretful, and keeps throwing his head seeking to escape the constant pressure on the bars.

The curb bit, in order to prevent the horse from taking a branch of the bit in his mouth, should be equipped with a lip strap that passes through a link in the center of the curb chain.

c. *Pelham and other type bits.*

The Pelham bit is a form of curb bit in which the snaffle ring is attached either loosely to, or as an integral part of, the branch of the bit. Pelham bits vary greatly in type and severity, but a sound rule is to use the mildest bit with which the horse goes well, while keeping his neck and head stretched out. A more severe bit will flex him and make control easier for a day or two, but grave difficulties usually will follow over-flexion.

The Pelham bit is adjusted in the same way as the curb bit. If the curb chain is passed through the snaffle ring of those Pelhams in which the snaffle ring is an integral part of the branch, it will not "ride up" under the horse's jaw and make it sore, but will stay in place in the chin groove.

There are many oddly-constructed patented bits and doubtless some of them are effective on some horses. In any event, if the riders think they are, it is a satisfying thought. Those ripe with experience usually conclude that there are no "get-rich-quick" methods of making a horse's mouth. Hackamores and other devices also may serve certain purposes, and perhaps aid in some phases of training, but when the pace and excitement run really high, a well-bred horse must know how to accept pressure on the bit without fear, flex, and come back to the rider's hand. To teach this, the only short cuts are found in the rider's skill and technique, which can speed training and the making of the mouth; or in those inherent qualities possessed by some horses—good conformation and sweet disposition. Spend the time necessary to train the colt and make his mouth correctly, and you will be repaid a thousandfold.

d. *Remedy for horses that put the tongue over the bit.*

When a horse has contracted the vice of putting the tongue over, or "balling it up" back of the bit, about the only solution is to tie the tongue in the mouth, making it impossible for him to pull it back from its proper position underneath the bit. A piece of cloth tape, about three-quarters of an inch wide, can be used for the purpose. Cut a transversal slit part way across the tape, and slip one end through the slit, forming a noose. The noose is put around the horse's tongue and drawn as tightly as possible without cutting off blood circulation. The two ends are then run down and tied underneath the lower jaw with enough tension to keep the tongue in its proper place.

Another method which is sometimes effective and which does not involve the actual tying of the tongue, is as follows: tie a piece of tape to the snaffle bit at the joint of the mouthpiece, having the two loose ends of equal length, before putting the bridle on the horse. The two loose ends come out of the corners of the horse's mouth and are tied to the two cheek-pieces of the bridle, or to the noseband at the middle of the horse's face. This latter method is quite effective. It works most easily if a small hole is punched in the noseband, running one end of the tape through the hole, and tying it to the other.

12. MISCELLANEOUS EQUIPMENT.

a. Noseband.

A noseband consists of a wide leather band, preferably adjustable, which is held in place by either a separate strap over the crown, or by loops in the cheek pieces of the bridle. It is used to prevent the horse from opening his mouth so wide as to escape the effect of the bit, to serve as a means for the attachment of the standing martingale, when this is used, or to enhance the appearance of the bridle.

The noseband should be adjusted so that it is two finger widths below the cheek bones, and permits the width of two fingers being inserted between it and the nose. Care must be taken that the band is not so low that the corners of the mouth are pinched between it and the bit.

b. Martingales.

Martingales are of two types, standing and running.

(1) A standing martingale sometimes called a "tie-down," consists of a noseband (described above), a heavy adjustable strap which runs from the noseband between the forelegs and fastens around the girth, and a light neck strap to keep the martingale from getting under the horse's feet when the head is lowered. It is used to prevent the horse from carrying his head so high that he escapes the correct effect of the bit.

The standing martingale should be adjusted so as to prevent the horse from raising his nose above the height of his withers. A standing martingale that is too long is useless; one that is too short cramps the horse.

(2) A running martingale is similar to the standing martingale, except that the former is not attached to the noseband. The heavy adjustable strap, which is Y shaped, has a small ring in each of the branches of the Y through which a snaffle rein passes. It is used for the same purpose as the standing martingale, and also to gain facility in turning a horse at fast gaits. It permits more freedom of movement than the standing martingale and may, therefore, be used in jumping.

A running martingale should be adjusted as follows: The snaffle reins are passed through the martingale rings, and the martingale is then adjusted in such a manner that when the horse's head is in normal position, the snaffle reins, when stretched from the pommel, will form a straight line from bit to pommel.

c. Breast plates.

The most satisfactory type of breast plate consists of a short, wide strap that passes over the neck in front of the withers, two adjustable straps that run from each end of the short strap back to the saddle, two adjustable straps that run down the shoulders to a ring on a breast plate, and another adjustable strap that runs from this ring and attaches to the girth after passing between the forelegs. This type is sometimes equipped with a strap that runs from the ring on the breast plate to the noseband, and acts as a martingale.

The breast collar is used for the same purpose as the breast plate. It consists of a wide strap that is adjusted to the girth on either side of the horse and held in place under the horse's neck by an adjustable strap passing over the neck. These devices are necessary on some slender-bodied horses and on horses which require some special security such as race horses to prevent the saddle from slipping to the rear.

The breast plate (or breast collar) should be adjusted as loosely as possible consistent with holding the saddle in place. The breast plate should be high enough to prevent the branches from galling the shoulders, but low enough not to interfere with breathing. Allowance should be made for motion and change of posture of the horse's neck.

d. Cavesson.

The cavesson consists of a heavy headstall and a hinged and jointed metal noseband that is heavily padded. The head-stall is similar to the ordinary bridle head-stall, except that it has an additional strap, the "jowl strap," that fastens snugly under the bulge of the lower jaw to prevent the cheek pieces being pulled into the horse's eyes. The noseband has a ring on the top for the attachment of the longe. The cavesson (with the longe) is used for exercising, disciplining and training horses—particularly young ones.

The cavesson is adjusted by buckling the noseband tightly just below the cheek bones. The lower it rests and the looser the adjustment, the more severe it is. The jowl strap should be buckled snugly under the lower jaw.

e. Longe.

The longe is a strong, light strap of webbing, or other suitable material, about thirty feet long. One end is attached to the noseband of the cavesson, with which it is always used. The longe should be folded in the hand in super imposed loops, like figures of eight; this arrangement will usually prevent its becoming entangled when played out rapidly.

f. Longeing whip.

A longeing whip has a stock about four feet long, and a lash six to eight feet long. It is used to keep the horse moving on the circle and to discipline him.

The longeing whip should be held in the same hand that holds the loops of the longe, that is, in the right hand when longeing to the left, and in the left hand when longeing to the right.

g. Hackamore.

The hackamore consists of a rope noseband, knotted under the horse's jaw, an ordinary head-stall, to hold this noseband in place, and a pair of reins. It has been used for many years on the western ranges in the training of cow ponies. It is an excellent device for controlling and training a horse without injuring his mouth, and is, therefore, used extensively in the early training of polo ponies. Other varieties of hackamores are made with metal branches that act similarly to the branches of the curb bit and exert pressure on the horse's nose by a lever action.

The hackamore should be so adjusted that it will rest on the cartilage of the horse's nose, about two finger widths above the upper edge of the nostrils. It should permit the passage of two finger-breadths between it and the branches of the jaw. Whereas, to be most effective, it should rest on the sensitive cartilage and care must be taken that it is not adjusted so low that it interferes with the horse's breathing or injures the sensitive cartilage.

CHAPTER IV
THE POSITION OF THE RIDER

13. To mount and dismount 18
14. The military seat 20
    a. General principles 20
    b. Position mounted 21
    c. Hands 27
    d. Balance 27
    e. Inclination of the upper body 28
    f. Summary 31
15. Manner of holding the reins 34
16. Posting 35

13. TO MOUNT AND DISMOUNT.
a. Mounting.
(1) *First Method.*

    The horse being saddled and bridled, the rider stands, half facing to the front, opposite the left stirrup. He grasps both reins in the right hand so that they come into the hand on the side of the forefinger which separates them, and so that the bight falls to the off side. He then places the right hand on the pommel and adjusts the reins so as to feel lightly the horse's mouth. Assisted by the left hand, if necessary, he places the left foot in the stirrup, and brings the left knee against the saddle. The rider should not press the left toe into the horse's side; to avoid doing this, he depresses the toe and keeps the knee firmly against the saddle. He now moves the left hand to the horse's crest, and, by a spring from the right foot, while keeping the left knee bent and still firmly against the saddle, raises the body erect in the stirrup. He brings the heels together for an instant, then passes the right leg, knee bent, over the horse's croup without touching it, and sits down lightly in the saddle. He releases his hold on the crest, places the right foot in the stirrup, assisted by the right hand if necessary, and takes the reins in one or both hands as instructed.

(2) *Second method.*

The rider stands, half facing to the rear, opposite the horse's left shoulder. He takes the reins in the left hand, with the little finger between them and the bight falling to the off side; adjusts them so that they give a gentle, even bearing on the horse's mouth, and places the left hand on the horse's crest. He then places the left foot in the stirrup, assisted by the right hand if necessary, and brings the left knee against the saddle. Without pause he places the right hand upon the cantle, and rises by an effort of the right leg aided by the arms. He keeps the left knee bent and firmly pressed against the saddle, the toe depressed, and the upper part of the body inclined slightly forward in order to keep the saddle from turning. He brings the right foot by the side of the left, changes the right hand to the pommel, passes the right leg, knee bent, over the horse's croup without touching it, and sits down lightly in the saddle. Assisted by the right hand if necessary, he then places the right foot in the stirrup and takes the reins in one or both hands as instructed.

b. Dismounting,

(1) *First Method.*

The rider takes the reins in the full grasp of his right hand so that the forefinger is between the reins and the bight falls to the off side. He then places the right hand on the pommel, adjusting the reins so as to lightly feel the horse's mouth. He moves the left hand to the crest, removes the right foot from the stirrup, and, with his weight on the left stirrup and his body inclined slightly forward, passes the right leg, knee bent, over the horse's croup without touching it. He brings the heels together for an instant, and then descends lightly on the right foot, which is followed by the left. He at once takes the reins over the horse's head in order to hold or lead him.

(2) *Second Method.*

The rider passes the reins into his left hand and places that hand on the horse's crest. He then places the right hand on the pommel, removes the right foot from the stirrup, and passes the right leg, knee bent, over the croup without touching it. He now changes his right hand to the cantle of the saddle and brings the right foot by the side of the left foot, the left knee being against the saddle and the upper part of the body inclined slightly forward.

He descends to the ground as described in the preceding paragraph.

14. THE MILITARY SEAT.
a. General Principles.
(1) *The correct military seat* permits the rider to remain master of his equilibrium, whatever may be the actions of his horse. It must be secure in itself and provide ease and comfort for both horse and rider. Such a seat is dependent upon *balance,* augmented by *suppleness, muscular control* of the body and *the use of the legs.*

(2) *Adaptability.*

The military seat, while obligatory in the Army, is also admirably adapted to all kinds of riding such as hunting, polo and jumping. For certain of these activities a different adjustment of stirrups may be necessary.

(3) *Importance of saddle.*

Without a properly constructed saddle, the deepest part of which is approximately in the center, it is extremely difficult to acquire or retain the correct military seat. The McClellan saddle is properly designed. Many flat saddles are too low at the cantle, or at the pommel. These faults place the deepest part of the seat of the saddle too far to the rear or too far to the front, making it difficult, and in some cases impossible, to assume the correct seat. Usually all issue flat saddles can be altered by changing the amount of padding at the cantle or pommel so that the deepest part will be correctly centered.

(4) *Principal elements.*

The principal elements entering into this discussion are the rider's *upper body,* his *base of support,* his *legs,* and his *equilibrium* or *balance.*

    (a) The *upper body* is considered to be that part of the body from the hip joints up.

    (b) The *base of support* is formed by those parts of the rider's body in contact with the saddle and the horse, from the points of the buttocks down along the inside of the thighs, to and including the inner knees and legs. The fleshy parts of the buttocks are forced to the rear and in no case form part of the seat.

    (c) The *leg* is that part of the limb between the knee and the ankle.

    (d) *Equilibrium* or *balance.*—It is quite evident that the rider, since he is constantly receiving impulses from the moving horse, frequently is in danger of losing his equilibrium and can retain it only by the clinging of the knees and thighs, reinforced by a sufficiently strong leg grip.

Fig. 2.

Correct seat with normal length stirrup-straps. Heels and knees forced well down; calves in close contact with horse; body inclined forward, distributing weight down thighs and in stirrups; loin and trunk erect; chest and head lifted.

Balance obviates the necessity for continuous leg grip, saves the legs from undue strain and fatigue, and is the principal requisite of a secure seat. Balance requires that the center of gravity of the upper body remain as nearly as possible over the center of its base of support. With the horse in motion, the center of gravity must be further advanced than when at the halt in order to compensate for the force of inertia which tends to over-balance the upper body to the rear and to leave the rider "behind his horse." When the center of gravity passes outside the limits of its base of support, the rider's balance is in danger of being lost and he must maintain it by gripping with his legs. A rider with a poor seat makes the grave mistake of pulling on the reins. *Balance must be entirely independent of the hands and reins.*

b. Position Mounted.

(1) *The rider sits with his crotch squarely in the center of the saddle, his weight distributed forward from the points of his buttocks into his crotch and down onto the inner thighs, knees and stirrups.*

(a) *At the halt,* the upper body, *due to a slight forward inclination from the hip joints,* is just in front of the perpendicular. Thus, its center of gravity is placed in front of the points of the buttocks. This facilitates the correct placing of the thighs and the proper distribution of weight.

(b) *When in motion,* to be in balance, the upper body is inclined farther forward from the hips. The lower thighs, knees and legs remain in close contact with the horse. The knees, ankles and heels sink at each stride, absorbing part of the shock and fixing the rider securely in the saddle. Inclining the upper body to the rear, or convexing the loin to the rear, places the center of gravity of the upper body in rear of the center of its base of support and causes the rider to sit on the fleshy parts of his buttocks. This faulty position tends to raise the thighs and knees, weakens the seat, concentrates the weight toward the cantle, and is unmilitary in appearance. It is fatiguing to the horse and often injurious to his back. The rider is said to be *"behind his horse."*

Fig. 3. Correct Seat at the Walk.

(2) *The thighs extend downward and forward, their inner sides resting without constraint on the saddle.* With the buttocks to the rear and the upper body inclined to the front as has been described, the thighs are naturally forced down, and the center of the saddle comes well up into the rider's crotch. The large fleshy muscles of the inner thighs are thus forced to the rear, and the flat of the thigh is permitted without muscular constraint to envelop the horse. Thus seated, a proper proportion of the rider's weight is distributed down his thighs, and the tendency to grip with them is avoided.

If the thighs are turned outward excessively, contact of knee and lower thigh with the saddle is lost, and the rider has neither the correct distribution of weight nor the proper base of support. Instability and lack of security result.

(3) *The knees are forced down as low as the adjustment of the stirrups will permit, without causing the stirrup straps to hang in rear of the vertical. Knees are neither limp nor stiff, nor is there normally any effort to "pinch" with them. Flexed and relaxed, they rest with their inner sides in continuous contact with the saddle.*

Properly placed thighs, as described in (2) above, naturally and correctly place the knees. Knees excessively turned out produce the same faulty results mentioned for similar incorrect positions of the thighs.

Knees excessively turned in force the heels out and cause the calves of the legs to lose proper contact.

Knees too high cannot form a proper part of the base of support. They place the rider behind his horse. Such a fault is an indication of the fact that either stirrups are too short, the rider is sitting back on his buttocks, or that the forward inclination- of the upper body is insufficient.

If knees are stiffened or straightened, the calves of the legs lose proper contact with the horse and the rider's seat is forced out of the saddle.

If the knees are limp, the legs go too far to the rear and the stirrup straps are no longer vertical. The heels come up and the crotch and buttocks slip too far forward in the saddle.

Thus, a faulty position of the knee is an indication of the fact that the entire seat is incorrect.

(4) *The legs, ankles, feet and stirrups are disposed as follows:*

(a) *The legs* extend downward and backward with the calves in light, elastic contact with the horse. The calves naturally fall into this position if the knees are flexed and relaxed. This contact of the calves is a means of communication between rider and horse and also assists security. When the legs are not in contact, communication is lost and their swinging confuses a well-trained horse; irritates a nervous one; and renders the seat insecure. Correct adjustment of stirrups assists materially in preserving leg contact.

(b) *Stirrup leathers are approximately vertical.* The length of stirrup is normal and approximately correct if the tread hangs opposite the lower level of the ankle bone when the rider is seated as described above, with his feet out of the stirrups and his legs hanging naturally, well down and around his horse. This is not a fixed rule, as the conformation of both horse and rider call for slight modifications.

Stirrup leathers for special forms of riding may be longer or shorter than described. For schooling, a longer stirrup should be used. For show jumping, steeple chasing, and racing, the stirrups should be shortened. Too long a stirrup diminishes the rider's base of support, renders balance from front to rear particularly difficult, and interferes with the proper use of the legs. Too short a stirrup raises the knees excessively, makes the seat insecure as to lateral reactions and causes undue fatigue when employed over long periods of time. With very short stirrups, unless the forward inclination of the upper body is increased materially so as to keep the center of gravity of the mass over the horizontal distance between the knees and heels (see Fig. 4), the rider is placed behind his horse with his weight towards the cantle of the saddle.

All requirements of military riding may be met by the normal adjustment of the stirrups as described above. Short stirrups should not be used except for the special purposes mentioned.

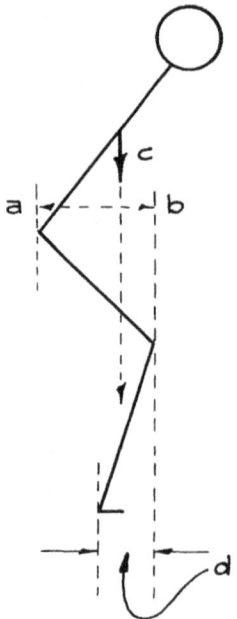

a-b - Represents horizontal length of base of support when seated.
d - Horizontal length of base of support when seat is out of saddle
c - Center of Gravity

Fig. 4.

The McClellan saddle is not suited for use with very short stirrups. Its high cantle prevents the buttocks from going to the rear as they must when stirrups are markedly shortened.

(c) *The heels are well down, the ankles flexed and relaxed.* The feet, turned out naturally, rest with the ball of the foot in front of the center of the stirrup tread. The rider normally supports the weight of his feet, legs and a portion of the weight of his thighs and upper body in the stirrups. This weight in the stirrups results from the forward inclination of the upper body, and not from "standing in the stirrups."

Ankles naturally break slightly to the inside, thus allowing the calves of the legs to rest against the sides of the horse. They should be relaxed, in order that the downward thrust on the stirrups will pass into the heels, causing them to sink below the level of the toes and allowing the ankle joints to flex freely with the movements of the horse. Ankles which are stiff cause the rider to carry the heels too high and result in unsteady legs and frequent loss of stirrups. They also restrict the rider in the proper use of his legs in the control and management of his horse.

Toes turned in stiffen the ankles, force the heels out, and cause loss of contact of proper parts of the calves of the legs. This fault reduces the security of the rider and makes the correct use of the legs impossible. Excessively turned out toes stiffen the ankles, put the knees out of contact, place the rider on the backs of his thighs and disrupt the seat.

The toes should not press down on the soles of the boots, but on the contrary, should be relaxed, thus aiding materially in obtaining a relaxed ankle.

It must be realized that when riding with stirrups, the calf muscles cannot be powerfully contracted except when the heels are driven well down. "Strong legs" and a strong seat are indicated by well depressed heels. They brace the rider against losing balance to the front and facilitate correct driving power against a stubborn mount.

(d) *Without stirrups the legs and feet hang in a natural manner,* except that the knees are flexed sufficiently to bring the legs into light, elastic contact with the horse in order to secure the seat and assist balance. The ankles are relaxed, thus permitting the toes to hang lower than the heels.

(5) *The posture of the upper body is that of the dismounted soldier at attention except for a habitual forward inclination from the hips.* This position distributes the rider's weight evenly over his base of support and so saves both horse and rider unnecessary discomfort and fatigue. If the upper body is inclined too far forward, the rider easily becomes unbalanced since at any violent

reaction his legs slide too far to the rear. On the other hand, the center of gravity of the upper body should never be so far to the rear as to be directly over the points of the buttocks.

(6) *The chest is lifted.* The shoulders are square without stiffness and carried in a plane at right angles to the long axis of the horse. Lifting the chest with the shoulders as described facilitates the maintenance of a correct posture of the upper body. Rounded shoulders cramp the chest, invite a general slumping of the back and loin and cause the elbows to fly out from the body. Shoulders forcibly carried back result in general contraction. Failure to lift the chest is often the cause of a poor seat, particularly in riding across country and over obstacles.

(7) *The head and chin are up, eyes looking to the front.* Due to its position and weight, the head has a great influence on the balance of the upper body. For this reason, it is important that it be correctly placed. If it is dropped down, the resulting tendency is to round the shoulders and back, thus destroying the ease and balance sought in the military seat.

The chin is held up without being thrust out or stiffening the neck.

The eyes are lifted, so that the rider may see where he is going. The bad habit of looking down at the horse's head and neck is dangerous for the rider and all others in the vicinity.

(8) *The arms, elbows, wrists and fingers* normally are relaxed, the elbows falling naturally in front of the hips. As long as the horse is going at the rate, gait and in the direction desired, only sufficient muscular energy is used to keep the forearms lifted to the proper position and to maintain the reins securely. A natural relaxation of the arms insures freedom and quietness in the use of the hands. Any involuntary contraction quickly communicates itself through the hands to the horse's mouth, resulting in loss of that calm confidence which the horse should always have in his rider.

(9) *The reins* are held in either or both hands, fingers softly closed. Backs of the hands are up and out at angles of about 30 degrees inside of the vertical.

(a) *The reins in both hands.*

The hands, with fingers relaxed, are separated about eight or ten inches and normally held slightly above the withers. The wrists are straight and supple, the forearms, wrists, hands and reins form almost straight lines from the points of the elbows to the horse's mouth. Some times, for corrective purposes, hands may be carried *higher* than just described. *However, they should never be carried any lower.* Hands carried too low give the impression of pushing down on the reins and cause the horse to seek to escape the downward effect of the bit on the bars of his mouth by raising his head and thrusting his nose into the air.

The elbows are carried slightly in advance of the points of the hips. Their position will vary from time to time in guiding or controlling the horse; but, with reins properly adjusted, they should never pass in rear of the hip joints.

When riding with a snaffle bit, maintaining a direct line from elbow to mouth is facilitated if the reins are taken into the hands between the third and fourth fingers, rather than underneath the little fingers.

(b) If only one hand is used, the free arm hangs naturally.

c. Hands.

Good hands are impossible to acquire without a good seat. Softness is an essential characteristic of good hands and must be developed. Relaxed arms, which permit the soft and elastic opening and closing of the elbow joints, combined with the lazy play of the wrists and fingers, give soft hands. As long as the horse is going at the speed and in the direction and manner desired, such hands smoothly follow the movements of his head and neck while maintaining soft, continuous contact of unvarying intensity with his mouth.

Rough hands are unsteady and quickly communicate unintentional impressions to the horse's mouth, making him nervous and difficult to control.

d. Balance.

When a rider so disposes his weight as to require the minimum of muscular effort to remain in his seat, and when the weight distribution interferes least with the horse's movements and equilibrium, the rider is said to be *"with his horse"* or *"in balance."* This condition of being "with the horse" is the keynote of riding.

When passing from the halt to motion, and when the horse is moving, the seat undergoes certain modifications. The rider must assume positions which assure his retention of balance and which keep him "with his horse." The knees, legs, ankles, and to a great extent the thighs, remain fixed in position. The upper body—the unstable part of the rider's mass—remains in balance over its base of support by appropriate variations in its degree of inclination towards the front, and thus overcomes the disturbing effects of the horse's movements.

Any change in the inclination of the body modifies the distribution of weight on the various parts of the base of support. As forward inclination increases, the rider's center of gravity is carried forward and downward. There is an increase in the weight borne by the knees and stirrups until finally when galloping fast across country or racing, and in certain phases of posting and jumping, the knees and stirrups support the entire load. Through the medium of the stirrup hangers [bars], this weight is distributed properly along the back of the horse.

e. *Inclination of the Upper Body.*
(1) *General.*

In forward movement, the degree of forward inclination of the upper body should vary with the speed of the horse and with the gait. This inclination always should be such that the rider remains in balance over his base of support. When the inclination of the upper body is not sufficient to maintain this balance, the rider is not *"with"* but is *"behind"* his horse. If it becomes excessive, the rider is not *"with,"* but is *"ahead"* of his horse.

The upper body is inclined forward from the hip joints. *The back should not break to the rear at the loin.* The eyes, chin, and chest are lifted in order that the back, while inclined to the front, may retain its normal posture, and the field of vision will be not reduced. To allow the back to break rearward at the loin and to permit the shoulders and head to drop forward, places the weight on the fleshy part of the buttocks and tends toward loss of balance to the rear. This in turn concentrates the weight of the rider at the rear portion of the saddle and, in marching, will result in undue fatigue and often injury to the horse's back in the region of the loins.

Suppleness, muscular control, and the opening and closing of the angles at the hips, knees and ankles supplement the inclination of the upper body and enable the skilled rider *to remain in balance with his horse.* In the case of unforeseen movements, such as shying or bolting, which tend to unbalance or unseat the rider, security is provided and balance retained, or regained, by an increased grip of the legs.

(2) *The upper body at the various gaits.*

(a) *Transitions. When passing from the halt to one of the various gaits* or when changing gaits or rates, the degree of inclination required of the upper body is dependent upon the suddenness of the change. In increasing gaits, the inclination must be sufficient to prevent inertia from carrying the center of gravity of the upper body in rear of the base of support. The change in inclination is made just prior to the change in gait.

(b) *At the walk* the upper body is inclined forward very slightly more than at the halt. As a result, despite the constant tendency to drift to the rear, caused by the horse's forward movement, the rider remains *in balance*. Thus seated, he neither slouches, concentrates his weight on the cantle, nor gets "behind his horse." The upper body has the same erect, alert appearance as that of the smart dismounted soldier at attention.

(c) *At the slow trot or trot (not posting), with stirrups,* the upper body remains practically erect without stiffness and has just sufficient forward inclination to keep its center of gravity over its base of support. Its forward inclination is approximately that assumed when at the walk.

At the slow trot, without stirrups, the beginner must lean back slightly in order to remain in balance. This is due to the change in position of the lower leg when riding without stirrups.

(d) *At the posting trot,* the rider's center of gravity under- goes more varied displacements than during any other gait. The length of his base of support varies from the maximum when he is in the saddle to the minimum when he is at the top of his rise. At this latter phase his base of support consists of his inner knees, legs and stirrups. *Sufficient forward inclination must be taken at all times to be in balance over the minimum base of support.*

The rider's body moves *forward* and slightly *upward,* then *backward* and *downward* in cadence with the beats of the gait. In rising to the trot, the angle at the hips should be opened as little as necessary, and the buttocks should remain to the rear. The upper body should maintain its inclination without appreciable change. *Excessively opening the angle at the hips during this movement causes the upper body to approach a vertical position and the buttocks to move too far to the front. These grave faults must be avoided.*

The upper body remains fixed in posture, and there is no sinking to the rear at the loin. Lifting the head and chest and keeping the hips and buttocks well to the rear produces an easy and natural posture. The chin is raised so that the plane of the face remains vertical. The rider sinks into the saddle very lightly on the upper thighs and crotch, and the points of the buttocks barely touch the saddle at each beat. The knee and hip joints serve as springs to make the reactions soft for both horse and rider.

Fig. 5.

Seated fully at the gallop. Note position of rider's boots and legs; feet braced against stirrups to prevent deranging of seat while legs gently urge horse ahead. Contact with mouth continuous, but very light, due to relaxed fingers, elbows, and shoulders.

A rider is said to be posting on the right diagonal when after rising, he comes back into the saddle at the instant the right forefoot comes to the ground.

(e) *At the ordinary gallop, when fully seated in the saddle,* the upper body is inclined slightly farther forward than at the walk or slow trot, but not as much as at the posting trot. The rider's thighs and crotch maintain continuous light contact with the saddle. At each beat of the gallop, that part of the rider's weight coming onto his thighs forces the *relaxed knees* downward, and they in turn transmit weight through the relaxed ankles into the heels. This automatically forces them down and causes the legs to maintain their proper position. The back and loin are straight without stiffness. The buttocks are forced well to the rear. The reactions of the gallop are absorbed by the play of the hip joints and *not by the relaxation of the loin.* Leaning backward at the gallop, or allowing the loin to break rearward, concentrates weight on the cantle and places the rider "behind his horse." He will then ride "heavily" instead of "lightly."

(f) *As the speed of the gallop is increased, the upper body is inclined farther forward from the hips.* The points of the buttocks are lifted clear of the saddle until the crotch is just out of contact. This places all the weight on the lower thighs, knees and especially in the stirrups. "Pounding" the saddle is eliminated, the rider is more comfortable, and the horse moves with more ease and freedom.

Rounding the back and loin entails loss of muscular control of the upper body and results in loss of balance. If balance is lost to the rear, the rider gets *"behind his horse"* and sits heavily, close to the horse's loins. Being *"behind the horse" makes* galloping laborious and painful to him and places the soldier in an unfavorable position for employing his weapons. When riding over-balanced to the front, the seat is insecure and the rider has difficulty in using his legs or hands to control his horse.

(g) *In decreasing rates and gaits, in halting and in backing* the rider must not lean back. If necessary, the forward inclination of the body decreases just sufficiently to enable the rider to remain in balance. As the horse decreases his speed or halts abruptly, the rider stiffens his back muscles and keeps his buttocks to the rear. He pinches momentarily with his knees and obtains a brace against his stirrups as a result of his low heels. These combined actions prevent him from losing balance to the front and permit him to remain off the cantle.

(h) *Summary.*

(1) Seat: Forward part of pelvic bones rest on saddle; crotch *well back* so as to be deep in throat of saddle; fleshy part of buttocks forced rearward toward the cantle at all times, and never allowed to slip forward under rider. Rider does not sit on buttocks but on inside of thighs and forward points of pelvic bones.

(2) Thighs: Flat; heavy muscles to rear of femurs; continuous contact down to, and including, inner sides of knees.

(3) Knees: Inside of knee bones are snugly against saddle skirts; pushed down as low as possible with stirrup leathers remaining vertical; not allowed to turn outward so as to leave space between them and saddle; normally do not grip tightly but sufficiently to keep whole thigh softly against saddle skirts. Knee joints almost completely relaxed, except when necessary to keep seat from being displaced forward or sideways as in stopping suddenly before an obstacle, etc.; must not be entirely limp or lower legs will slip too far to rear, heels come up, seat slide forward, and rider hump his back.

(4) Lower legs: Inner and upper portions of calves always remain in soft contact with sides of horse; no great effort required to keep them there. Calves squeeze to drive horse forward, and to maintain seat in case of emergency. In latter case, knees also increase grip. When stirrup leathers are vertical, knee joints relaxed, and heels thrust down to absolute limit

permitted by relaxed ankle joints, position of lower legs is correct. Heels should be kept down when squeezing or gripping with calves. Spurs are used just in rear of girth—not far back against the flank.

(5) Ankles: Habitually relaxed, allowing weight transmitted down thighs, through partially relaxed knees, to sink into heels of boots. If trunk is correctly inclined forward at the hips, portion of rider's weight necessarily runs down through thighs and automatically flexes ankles and drives heels down.

(6) Feet: Turned out so that upper, inner sides of calves rest against horse. Toes make angle of twenty to forty-five degrees with longitudinal axis of horse. Stirrups slightly in rear of the ball of the foot, permitting all weight to sink into heels. For schooling, balls of feet may rest on stirrup treads.

(7) Heels: Thrust far down. Give brace to feet against stirrups if horse, checks suddenly, so seat cannot slip forward. Receive all weight coming into stirrups if feet and ankles are correctly relaxed.

(8) Trunk and Loin: *Carried in same posture as when standing erect in the "Position of a soldier' except that whole trunk is inclined to front from hip joints.*

When fully seated in saddle at halt, walk, slow trot or canter, center of gravity of the trunk falls just in front of pelvic bones. *At posting trot or gallop,* center of gravity is approximately over knees, and the trunk's forward inclination is greater. Loin is habitually "hollowed out" in its normal, natural position; never remains convex to rear. Buttocks should be well in rear toward cantle of saddle, but due to body's forward inclination *no weight is on them.* Buttocks provide rear counterbalance for the forward-inclined trunk. Knees, over which trunk and buttocks are balanced, are center of motion when posting, galloping or jumping.

(9) Chest, Head and Chin: Lifted. *Whole body carried lightly. There should be a feeling of stretching the spine up ward and making the body tall.*

(10) Length of Stirrups: After some slow *trotting without stirrups* (when seated as prescribed above for slow paces) and with legs hanging down in natural position by horse's sides, treads of properly adjusted stirrups should hang even with center of large bones on inner sides of ankle joints. A little variation in length is inevitable due to difference in conformation of both people and horses. For fast cross-country work or jumping, stirrups are shortened from one to four holes. The shorter the stirrups are adjusted, the greater must be the inclination of the upper body.

(11) Important Points:

a. Any faulty attitude of one part of the rider will cause faults in other parts, thus throwing the whole seat out of adjustment.

b. When the horse checks suddenly, goes down a steep incline, or lands after a jump, the knee joints *should stiffen* a trifle, and in conjunction with the lower heels, permit the feet to brace the whole body against the stirrups. The knees also should grip the saddle more tightly and the back muscles should stiffen in order to keep the spine straight. These actions prevent the body's toppling forward, and hold the seat secure.

The above actions are easily accomplished after practice, and also serve to prevent the buttocks' slipping forward and the lower legs going to the rear which disrupt the whole seat and involve surrender of balance and control of the horse. *If the heels are thrust down and the back is kept swayed, the forward inclination of the body, even when checking the horse very quickly, can, and should, be maintained.*

The rider's knees, when in the position described, are approximately in the transverse vertical plane containing the horse's center of gravity. Hence, with the rider correctly seated, their centers of gravity fall approximately in the same vertical line. During movement there is, of course oscillation of the centers of gravity of both, but they remain approximately in the same vertical line if the rider is constantly "with his horse."

c. Stiffness should be avoided. As much relaxation should exist throughout the whole anatomy as is consistent with maintaining muscular control of the body, balance, and the seat steadily in place.

(12) Position at Walk and Trot: In order to lessen fatigue to the horse *it is absolutely vital to maintain slight forward inclination of the body at the* walk and trot. This keeps the rider's weight distributed down his thighs; whereas leaning backward or sitting bolt upright, concentrates it far back on the cantle, which is very tiring to the horse.

(13) Liberty of Head and Neck: In addition to using the correct seat every effort should be made to allow horses *maximum liberty of head and neck.* At the walk in particular, the reins should be very long permitting the horse to stretch his head and neck into a low, extended position favorable to long strides and comfort. The hands remain still at the trot; at the gallop they move back and forth with the horse's movements, "following the mouth." Elbows must be partly flexed so as to be soft and elastic.

(14) How to Test Correctness of Rider's Position: If the rider is in balance as a result of his upper body being properly inclined forward, he is able at the walk, trot or gallop, *without first leaning farther forward* and without pulling on the reins, to stand in his stirrups with all his weight in his depressed heels.

In executing this exercise the seat is raised just clear of the saddle by stiffening the knees but keeping them partly flexed. The upper body remains inclined forward at hips. At the trot one hand should touch the horse's neck *lightly* to assist in remaining in balance. At the walk or gallop, the rider, if his seat is correct, should be able to stand in his stirrups without the aid of his hand. A rider who can execute the above exercise at all gaits and without first changing inclination, is in balance and never "behind his horse." The majority of those *not* in this position partly maintain their balance by hanging onto the reins, thus unnecessarily punishing their horses' mouths as well as their backs.

15. MANNER OF HOLDING THE REINS.
a. Single Rein.
With a single rein, such as generally is used with a snaffle bit alone, the reins enter the two hands between the little and third fingers, passing up through the palms, and out over the index fingers. The thumbs are placed on top of the reins, pressing them against the middle joints of the index fingers, which prevents the reins from slipping.

When both reins are held in one hand, the rein from the empty hand enters between the ring and middle fingers, or between the middle and index fingers, as desired. The two bights, (loose ends of the reins), then pass over the index fingers, and are held in place by the thumb, as in the case when the reins are held in the two hands.

b. Double Reins.
If the reins are held in both hands, when using a double bridle, (one having both the curb and snaffle bits, often called "bit and bridoon"), the snaffle rein enters each hand underneath the little finger, and the curb rein enters between the little and ring fingers. The two reins run up together through the palm and out over the index finger, and are held in place by the thumb.

If all reins are held in one hand, one curb rein enters between the little finger and the ring finger, the other between the ring and middle fingers; one snaffle underneath the little finger, the other between the middle and index fingers. The bights pass out of the palm over the index finger, and are held down by the thumb. Thus the snaffle reins are always outside the curb reins, relative to the horse's neck and to the fingers.

There are numerous ways of holding the reins, many of which may be satisfactory. However, those just given are thought to be as good as, if not better than, any of the others.

16. POSTING.

a. Posting, or rising to the trot, has been quite generally adopted as a means of reducing the shock of this gait for both horse and rider. It is accomplished as follows: As the horse takes up the trot, the rider inclines the upper part of the body forward, supports himself by pressing the knees against the saddle, and then permits himself to be impelled upward by the thrust of the horse's hind leg (the left, for example). He remains up during the stride of the right hind leg, and sits down just in time to be impelled upward again by the next thrust of the left hind leg. He continues in this way, avoiding the alternate thrust of the hind leg.

b. When learning to rise to the trot the beginner will make more rapid progress if, at each effort to rise, he strokes the horse's neck. Stroking the horse's neck assists the rider to catch the rhythm of the motion; it also causes him to incline the body forward at about the correct angle. In rising to the trot, the buttocks should be raised moderately from the saddle; contact with the saddle should be resumed gently and without shock; the knees should be pressed snugly against the saddle; the lower leg should be kept perfectly still, the ankle-joint supple, and the heel slightly lower than the toe. The rider's head should be up, and his eyes glancing well out to the front; he should not lower the head and look down toward the horse's front feet.

c. When the rider sits down in the saddle each time the right fore foot strikes the ground, he is said to be posting on the right diagonal; when he sits down each time the left fore foot is planted, he is posting on the left diagonal. The rider should frequently alternate diagonals in order to insure equal development and power in the hind legs of the horse. On straight lines it is immaterial which diagonal he posts upon; provided he uses both diagonals equally; but in a riding hall or inclosure it is essential that a fixed rule be insisted upon.

d. A horse when worked in a hall travels a great deal of the time on a curve in such a manner that his outside lateral travels a greater distance than his inside lateral, his outside hind leg further than his inside leg. On the left hand, for example, the rider should post on the outside or right diagonal, receiving the thrust of the left hind leg, which has the shorter distance to travel, and thus equalizing the work of the hind legs. The converse is true if the horse is on the right hand.

CHAPTER V
SUPPLING EXERCISE

17. Purpose 36
    a. Establishing the confidence of the rider 36
    b. Precautions 37
18. Exercises to Supple the Rider 39
    a. To supple the neck 39
    b. To supple the shoulders 39
    c. To supple the loins 40
    d. To supple the hip joints 41
    e. To supple the knees 42
    f. To supple the ankles 42
19. General Remarks 42

17. PURPOSE.

a. Establishing the Confidence of the Rider. The mounted instruction of the inexperienced rider is hindered at the beginning by an unreasonable and instinctive fear of the horse, over which he feels he has no control, and a resulting revolt of his nervous and muscular systems which leads to contraction. This fear is overcome by establishing the confidence of the rider in himself and in his horse by means of mounted gymnastic and suppling exercises.

The particular contractions which are present at the beginning of individual work will disappear with the practice of suppling exercises. In order not to lose any of the useful effects, a logical order of application must be followed beginning with the neck, then the shoulders, loins, and hip joints. Movements of the thighs and legs must not be undertaken until ease in maintaining the correct positions and balance of the body has been obtained.

In attempting to obtain suppleness it should be remembered that good humor is conducive to relaxation, which in turn leads promptly and directly to confidence.

b. Precautions.
(1) The suppling exercises, like all physical culture exercises, depend for their good results upon the regularity and thoroughness with which they are practiced daily. Thus, long hours or intensity of application on one day do not compensate for the lack of practice on previous days. Furthermore, to obtain the maximum result from these exercises, they must be performed correctly. Performed incorrectly they may serve only to confirm an existing fault. Hence, it is highly important that the instructor and the student alike have well in mind the object in view in each instance. Accordingly, the instructor should frequently gather the class about him and briefly explain and illustrate an exercise. Immediately following this he should cause the class, still standing at the halt, to practice accurately the exercise just explained. When understood by the class and systematically applied by the instructor, the suppling exercises produce quick results. By means of them the rider is confronted with little problems in mental, physical and nervous control, the solution of which can come only through continuous daily practice.

(2) The gaits employed during the suppling exercise should be the walk, the slow trot and the gallop. The gallop should be employed very early in the instruction, and used chiefly thereafter; it is not too rough and is, moreover, the most favorable gait for suppling the rider's loins. The gallop can be maintained for long periods without undue fatigue. The slow trot is, on the contrary, very trying. With inexperienced riders it produces great fatigue, and if prolonged, is liable to cause such abrasions of the skin and soreness of the muscles as to retard true progress rather than promote it. There should, nevertheless, be much work at the slow trot, but in short periods only, and interspersed with a variety of other and less difficult exercises.

(3) The instructor causes the riders to take the track without regard to distance; if they march to the right hand, they hold the reins in the left hand; if to the left hand, they hold the reins in the right hand. Thereafter they take the reins in one or both hands, drop and retake them, as necessary, without command.

(4) Occasionally, as a temporary measure, the instructor may cause the riders to knot the reins and release them entirely. Habitually, however, the reins should be held while doing the suppling exercises; the rider is then cultivating a good hand in conjunction with a good seat. Moreover,

the instructor is then able to point out instances illustrating the effects produced upon the bridle hand by the stiffness and bobbing of the body due to a poor seat.

(5) While practicing an exercise, the rider should pay strict attention to the position and steadiness of the bridle hand and the suppleness of the wrist, elbow and shoulder. He should endeavor to coordinate and to separate his forces; thus, while moving or using one part of the body, he should not unconsciously or nervously contract the muscles in an unrelated part of the body or move the bridle hand.

(6) For the instructor to merely advise the riders to relax is wholly insufficient; the instructor corrects deficiencies by employing the suppling exercises. He observes the needs of individual men and constantly places before them, for their solution, a variety of little problems in physical coordination and self-control.

(7) The riders being suitably disposed on the track at the gait ordered, the instructor places himself near to and facing the track and gives the command for the desired suppling exercises. The movement is then begun at once, individually by the riders. In like manner, at the instructor's command *as you were*, each rider discontinues the exercise. The exercises are not performed in unison.

(8) The instructor, in his observations and criticism, confines his attention chiefly to the riders passing in front of him. Those riders who are beyond the view of the instructor are then enabled with less self-consciousness to recover their equilibrium and to adopt corrective measures. They are not, however, to cease to perform an exercise, even though beyond the view of the instructor, until the command is given to do so.

The instructor's commands, observations, and criticisms should be expressed in short sentences, employing familiar phrases easily understood, the voice loud enough to be heard, but without shouting and scolding. When prolonged or special explanations are necessary, the rider in question should be required to fall out of the column and report to the instructor.

If several riders are involved, the instructor should cause the whole column *to ride in;* thus grouped about him, all profit by the instruction given.

(8) The suppling exercises engage the rider's attention and lead him to ride without conscious effort, thereby bringing about relaxation of the muscles. With this in view, instructors may, later on, add other suitable exercises for the purpose of varying the work and adding to its interest and enjoyment.

18. EXERCISE TO SUPPLE THE RIDER.

The suppling exercises have for their object the assistance of the rider in developing the strength, pliancy, and easy control of the parts of the body most affected in riding. These are the neck, shoulders, loins, hip joints, knees, and ankles.

a. To supple the neck.—Object: To combat rigidity of position due to contraction of the muscles in the region of the neck.

(1) *Rotate the neck.*

Each rider slowly turns the head and eyes as far as possible to the right and then, without pause, turns them as far as possible to the left.

(2) *Flex the neck.*

Each rider slowly carries the head a far as possible upward and backward and then, without pause, as far as possible forward and downward.

Cautions: Do not derange the position of the shoulders and arms.

Do not contract the muscles of the neck, back, shoulders, and arms.

Do not move the bridle hand.

b. To supple the shoulders.—Object: To combat rigidity of position due to contraction of the muscles in the region of the shoulders.

To cultivate coordination of muscular effort; that is, to suppress the involuntary movement or use of one member of the body while moving or using another member.

(1) *Rotate the right arm vertically.*—Each rider extends the right arm straight upward, elbow, wrist and fingers relaxed, palms of the hand to the front and then, without pause, carries the arm to the rear, downward, forward, and upward in such a manner as to describe, with uniform movement, a vertical circle.

(2) *Strike blows to the front.*—Each rider, with closed fist, strikes a blow straight to the front, withdraws the arm and continues to strike blows.

Cautions: Do not derange the position of any other part of the body.

Do not contract unduly the muscles of any part of the body.

Do not do the exercise in a constrained manner.

Do not interrupt in the slightest, the smooth and easy continuity of the movement; to do so is an indication of involuntary muscular contraction due to nervousness.

Do not move the bridle hand.

c. To supple the loins.—Object: The exercises indicated in (1), (2), (3), and (4) following have for their object to give the rider practice in *rotating* the body in the region of the loins; those indicated in (5), (6), (7), and (8), practice in *flexing* the body in the region of the loins. The movements should be executed in such a manner as gradually to bring about the desired result, which is suppleness. Nothing is gained by exaggerated positions; they often cause strains of the muscles and tendons or involuntary nervous reactions which produce constraint and stiffness of the body rather than suppleness.

(1) *Rotate the right arm from front to rear, horizontally.* Each rider extends the right arm straight to the front, elbow, wrist and fingers relaxed, palm upward. He then, without pause, and by rotating the body at the region of the loins, swings the shoulders and arm to the right, the arm moving in a horizontal plane until the hand is over the horse's croup, the eyes following the hand. Each rider thus continues the exercise, swinging the shoulders and arm in a horizontal plane, by rotating the body at the waist.

(2) If the riders are riding without holding the reins, the instructor may command: *Rotate both arms from front to rear, horizontally.* Each rider extends the arms to the right and left, respectively, to a horizontal position, palms up, elbows, wrists and fingers relaxed. He then, without pause, and by rotating the body in the region of the loins, swings the arms to the front and rear in such a manner that when the right hand is over the horse's neck the left hand is over the horse's croup, and vice versa.

(3) *Stroke the horse on the left shoulder with the right hand*—Each rider, by rotating the body in the region of the loins, carries the right hand over the reins and downward to a position opposite the left shoulder of the horse. He then gently strokes the horse's shoulder.

(4) *Stroke the horse on the left shoulder and the right haunch with the right hand.*—Each rider, in the same manner as in the preceding exercise, strokes the horse on the left shoulder with the right hand and then, without pause, while rotating the body in the region of the loins, turns the shoulders and arm to the right rear, the eyes following the hand, and strokes the horse on the right haunch. He thus continues the exercise, alternately stroking the horse on the left shoulder and right haunch, by rotating at the waist.

(5) *Stroke the horse on the right shoulder with the right hand.*—Each rider, by flexing the body at the waist, leans slightly forward and downward and gently strokes the horse on the right shoulder.

(6) *Stroke the horse on the right flank with the right hand.* Each rider, by flexing the body at the waist, leans slightly backward and downward and gently strokes the horse on the right flank.

(7) *Swing low to the front with the right hand.*—Each rider extends the arm vertically upward, palm to the front, leans to the right and downward, by flexing the body at the waist, and makes a low, sweeping stroke from the rear to the front with the right hand and then resumes the initial position of the body, the arm extended vertically upward.

(8) If the men are riding without holding the reins, the instructor may command: *Stroke the horse on both shoulders with both hands;* or, *Stroke the horse on both flanks with both hands;* or, *Stroke the horse alternately on both shoulders and both flanks with both hands.* The manner of executing each of these exercises is obvious.

d. To supple the hip joints.—Object: To cause the thigh bones to lie flat against the saddle.

Rotate the right thigh. Being at the halt, each rider removes the thigh from against the saddle, straightens the leg and carries it back until behind the vertical. He then rotates the thigh at the hip joint by turning the knee inward and, pressing the thigh hard against the saddle, drawing the thigh forward to its normal position. The friction of the thigh against the saddle presses backward the muscles underneath and back of the thigh, thus permitting the flat of the thigh to rest solidly against the saddle. The lower leg is replaced in its normal position.

The foot should hang naturally, without constraint. No attempt should be made to turn the foot either outward or inward. Its position in this respect is dependent upon, and a natural consequence of, the flatness of the inner surfaces of the thigh and the degree of its rotation inward.

Cautions: A long-legged, lean man, wide in the crotch, needs little of this exercise. A short legged, fat man, narrow in the crotch, needs much of it; his muscles resting like a soft cushion between the thigh bone and the saddle, seriously handicap him in acquiring a good seat.

This exercise should be given at the halt, habitually; if executed while the horse is in motion, serious injury may occur in the region of the hip joint.

The riders being at the halt, the instructor directs them to rotate the right thigh and carefully fix it in position; next, similarly, the left thigh. He then causes the riders to take the walk, the slow trot, or the gallop, during which they should try out and endeavor to confirm the new position. After a few moments, however, the riders will become shaken out of the exact position assumed. The instructor then brings them to the halt, directs them to again rotate one thigh, then the other, and to fix their position, after which he causes them again to move forward. And so on, until the exercise is discontinued.

e. To supple the knees.—Objects: To render the knee joint supple.

To combat the involuntary contraction and stiffness to which the knee joint is liable.

To cultivate coordination of muscular effort. *Flex the right or left knee.* Each rider, removing the lower leg from against the horse, carries it backward and upward until horizontal and then, without pause, returns it to the normal position.

Cautions: Do not move the leg with a jerky motion.
Do not contract another member of the body.
Do not contract the muscles of the loins.
Do not derange the position of the upper body. Do not move the bridle hand.

f. To supple the ankles.—Objects: To remove involuntary contraction and stiffness from the ankle joints.

To assist the rider to lower the heels properly when riding with stirrups.

To assist the rider to retain his stirrups by the easy flexibility of the ankle joints.

To assist the rider not to stand stiffly in his stirrups.

(a) *Rotate the right (left) foot.*—Each rider traces with the toe of the right foot, by a slow and uniform movement, a circle in a vertical plane perpendicular to the side of the horse—clockwise with the right foot, counter-clockwise with the left foot.

(b) *Flex the right ankle.*—Each rider slowly flexes the right ankle by raising the toes upward, by a slow and uniform movement, as far as possible and then, without pause, extending them downward as far as possible.

Cautions: Do not, by the rotation of the feet, derange or affect any other part of the body.

Do not move the feet with a jerky motion.

19. GENERAL REMARKS.

After the men have learned the suppling exercises described, the instructor may, in his judgment, introduce such others as will add interest and variety to the work. On the whole, however, it is best to follow daily the exercises as prescribed. It is by close attention to little details and by oft-repeated practice extending over a long period of time that the rider gains complete control over his nervous reactions, acquires a sense of balance at the faster gaits, and learns to exert his strength easily, accurately and in the required measure. It is for the instructor, by exacting close attention to and causing continual practice of these exercises, to afford the means by which all the men may reach the desired end, which is regularity and security of the position mounted, suppleness of body, and coordination of effort in the exertion of strength.

CHAPTER VI

THE APPLICATION OF AND ACTION OF THE AIDS

20. Object..44
21. Equilibrium 45
    a. Direct equilibrium......................................45
    b. Lateral equilibrium 46
22 The natural aids..46
    a. Action of the legs 46
    b. Action of the reins....................................49
        (1) Opening rein 52
        (2) Direct rein..54
        (3) Bearing rein 56
        (4) Indirect rein of opposition
        in front of the withers................................58
        (5) Indirect rein of opposition
        in rear of the withers 60
    c. The weight 62
23. Accord of the aids..63
24. Lateral and diagonal aids................................64
25. The artificial aids...65

20. OBJECT.—The object of the study of the application of and action of the aids is to teach the rider the use of the means at hand to control the horse at all gaits, in all directions, and over any terrain.

To manage the horse is: To put him in movement. To regulate that movement. To direct that movement. For this it is necessary:

To know the means that nature, art, and science have put at the disposition of the rider (study of the aids). To employ these means (use of the aids). To harmonize these means (mastery of the aids).

21. EQUILIBRIUM.

a. Direct equilibrium is that in which the weight of the horse and the rider is carried neither to the right nor the left.

Experiments have shown that the weight of a mounted horse, halted and straight, is so distributed that the forehand is charged slightly more than the hind quarters. In addition, the horse is so constructed that the front legs have a role especially of movement, while the hind legs have that of propulsion.

This double action, which places the center of gravity near the shoulders and removes it from the hind quarters, greatly favors movement. In fact, the more the center of gravity is advanced, the greater the forward effort of the hind quarters in movement to the front. Moreover, this displacement of the center of gravity tends to cause the horse to move from his base of support, so that the forces of gravity draw the mass forward at the same time that the hind quarters push it in the same direction. In this the leg movements of the horse are extended. Inversely, if the center of gravity moves to the rear and is just above the points of support of the hind quarters, the latter exert their force from below upwards and the leg movements of the horse gain in height. In the case where the equilibrium is over the shoulders, the hind quarters control very slightly the mass, which they can only push forward and which is itself constantly drawn forward by its weight, like the body of a man in a bent forward position. In the other case, where the equilibrium is over the haunches, the hind legs support the weight and are able to move it freely in all directions. In this the movements are less extended, but the horse is more easily handled.

These very simple considerations must be understood by all riders, as they are the basis of the methods used in all demands made of the horse. From them the following conclusions are deduced:

(1) When the horse desires to advance, he places himself in the most favorable position for movement. Consequently, he displaces his weight toward his shoulders by extending his neck and head which serve as his balance.

(2) It follows, therefore, that in order to move forward and to increase the gait, the rider should allow the horse to extend and lower somewhat the head and neck. To decrease the gait, however, they must be raised.

(3) A horse can be ridden accurately and agreeably only if he is at all times ready to move forward, for in this case only does he place himself in accord with his conformation, which is essential to free and prompt obedience to the demands of his rider. He is then said to have impulsion, without which he can no more be controlled than can a boat at anchor be controlled by its rudder.

Moreover, as the horse's role is to carry his rider from one point to another, he does not accomplish this unless he is at all times ready to move forward.

Because of these considerations, we strive to maintain the natural impulsion of the horse, or to produce it in those which lack it. The legs must produce or awaken this impulsion, and the hands, which receive and regulate it, must avoid contradicting their action. This means, as we will see later on, in never pulling on the reins.

(4) The rider is able to obtain easy control of his horse, and the latter to submit to such control, if the methods employed conform to the physical characteristics of the horse.

b. Lateral equilibrium is that in which the horse is placed in order to weight a shoulder or haunch or an entire side more than the other. It is used for changes of direction, parallel displacements, etc. It is evident that if a horse in movement is forced to carry the weight of his forehand to one side, the entire forehand tends to move in that direction. This displacement may be said to be obligatory if that of the center of gravity is accentuated. Since the change of direction is the result of the displacement of the forehand, the best means to employ it in order to obtain this result is to cause the reins to act after moving the hands toward the side towards which the movement is desired, and not by rearward traction of the rein on that side.

22. THE NATURAL AIDS.

In equitation everything depends on controlling the equilibrium. This is true in training and also in the practice of equitation. The rider obtains control of the equilibrium by means of the natural aids, which are the legs, the reins, and the weight. These are assisted at times by the spurs, whip, and voice.

a. Action of the legs.

The most important role of the legs is, in acting simultaneously, to produce or maintain impulsion without which the horse is of little use. They put the horse in movement and obtain increases in the gait. Equipped with the spur, they are valuable aids to the rider in obtaining obedience to his will. It can be seen, therefore, how important it is to constantly perfect their use, for used improperly their effectiveness and authority is diminished, and the rider is unable to obtain as he should the impulsion which the legs are the best means of producing.

In addition to acting simultaneously, the legs often have another role which, while comparatively less important, is very useful, viz: using one with greater force than the other. They are thus used to displace the hind quarters laterally, or to carry the weight on to a certain haunch.

(1) *Equal action of the two legs.*

As we have just seen, the purpose of this action, and the result with a trained horse, is to obtain forward movement, increased impulsion, or the acceleration of the gait. To obtain this result, the legs may act by a simple pressure of the knees, or by pressure of the knees and calves. The first method is good only with sensitive horses; with the majority the second method is used. The effect produced is greater when the pressure is more energetic and produced in rear of the cinch. Habitually, the leg causes sufficient pressure if applied on or slightly in rear of the cinch. If the pressure is not sufficiently effective, it may be employed slightly further to the rear. Perfection is obtained when the movement of the legs is almost imperceptible and their effects varied only by slight changes in the degree of pressure. In the case of a horse which does not respond sufficiently to these demands, the leg, with the knee bent and the heel low, must move further to the rear. Under these conditions, more energetic measures should be resorted to. The first method employed consists of mild and repeated taps of the calves of the leg, continued until the desired result is obtained, when their action should cease and not be repeated until such action is again necessary.

If light taps are ineffective, blows with the leg should be resorted to. This consists of removing somewhat the calf of the leg and bringing it against the horse strongly enough to produce the desired result. In this movement the knee should not be raised or moved from place, in order not to derange the seat or hands. If blows with the leg are insufficient, this method should not be continued. Rather than such repeated action, resort should be had to brief and energetic action with the spurs.

It is considered essential to repeat here the fact that when the legs act to produce impulsion, *the hands should yield thus permitting the horse to obey the demands of the legs, and not, as is only too natural with young riders, act with greater force.*

The frequent fault of continued action of the legs when the horse is prepared to move or when he has obeyed the movement demanded, should be avoided, as in such cases the impulsion is increased in a way prejudicial to the result desired, and the hands have to act in opposition to the impulsion wrongly ordered by the legs. *The repeated countermanding of this increase in impulsion as soon as it is demanded, soon results in the failure to obtain it in spite of the demands of the legs, whose effectiveness is soon decreased.*

(2) *Unequal action of the legs.*

When one leg acts with greater force than the other, the haunches are displaced to the opposite side and the horse is said to range his haunches. This effect is often useful in preventing a horse from side stepping, in straightening him, or in changing direction in a cramped space, etc. Its greatest use is in permitting the rider to traverse and two track his horse, which movements are as useful in training the rider as they are in suppling and disciplining the horse.

The lateral action of the leg is exerted in accordance with the methods previously described for the equal action of both legs, taking into consideration the placing of the legs and the force with which they are used. This will be dealt with in detail in the discussion of traversing the horse and of work on two tracks.

(3) *The spur*

The rider should not wear spurs until his seat is sufficiently secure so that involuntary displacements will not cause him to employ them uselessly. As long as the pressure of the legs is sufficient to obtain and maintain the impulsion desired, it is a poor policy to use the spur. It should only be made use of when the action of the legs is ineffective, and then in a brief, but more or less vigorous and repeated manner, proportioned to the resistance encountered.

This point must be insisted upon, for if the spur remains in contact with the horse, the continuity of the pain or sensation may cause resistances which, being of a physiological nature, are too much for the animal who then disobeys his rider and defends himself. Therefore, keeping the spur in continuous contact with the horse's flanks must be avoided and such action replaced by repeated applications proportioned in amount and intensity to actual needs.

The length of the spurs should vary according to that of the stirrups and legs of the rider, the form of the horse, etc. For a given rider and horse, this length must be such that it can be used readily, without involuntary action. They should be well adjusted so that the rider can be certain of their action.

b. Action of the reins.

The reins are an intermediary between the hand of the rider and the horse's mouth. The bars, which are their point of application, are very sensitive, and that which was stated as to the necessity of having the horse responsive to the legs, may well be repeated here, for if the legs produce the movement of the mass and the displacement of the center of gravity, it is the hands which regulate the latter in order to establish general equilibrium. Finesse in equitation lies in the tact which the rider uses in applying his aids.

The manner in which contact between the mouth and the bit is established greatly influences the training of the horse as well as the equitation of the rider.

In many cases it is with great difficulty that the rider is able to make his horse understand that impulsion is always demanded by the action of the legs.

Impulsion may be lost, and the horse put behind the bit and upset, by improper use of the reins. In order to avoid this disastrous result, the reins should act by the effect of the impulsion produced by the legs. The use of the reins, therefore, instead of hindering the impulsion, becomes a result thereof, using, controlling and developing it.

To apply this principle *the horse must be pushed up onto the bit and not the bit pulled back to the horse.*

This is obtained as follows: Through the action of the legs, the horse who has impulsion, stretches out his neck to begin or to accelerate the forward movement. If the fingers are closed at this moment, the extension of the neck produces a stronger contact of the mouth with the bit, causing the latter to act.

The action of the reins is thus produced by the effect of the horse's submission to the legs and by profiting from the impulsion produced.

This method also has the advantage of not provoking resistances to the hand as happens when one pulls on the reins, because in the latter case, the action of the bit contradicts that of the legs instead of being the result thereof.

Finally, when the reins act they find all the powers of the horse gathered and ready to displace his mass at the slightest indication.

Numerous difficulties arise if the reins pull on the horse's mouth, instead of acting as above stated. In effect, in pulling, they act either alone or at the same time as the legs. In the former case they find the horse inert and without impulsion, and must act against the weight of the mass. The horse, instead of moving himself, allows his center of gravity to be displaced by their action. He is then heavy on the hand and managed with difficulty, and he can take advantage of this to resist the rider's demands. On the other hand, if the legs act at the same time that the reins act on the horse's mouth, these aids contradict each other, for the neck is drawn to the rear at the very moment it should seek to extend itself under the action of the legs. Acted upon by two opposite forces, the horse is forced to disobey one in order to obey the other, provided he does not avoid them both.

For effects to be exact, the reins must remain adjusted and stretched during work; if they are flapping, the indications of the hand will probably not reach the horse, or if they do, they will arrive confused, or in the form of brutal and awkward jerks. Contact is that gentle liaison which should exist between the hand of the rider and the mouth of the horse. This soft, steady contact with the mouth is maintained by having the elbows partly bent, and the joints of the shoulders, elbows, wrists and fingers almost completely relaxed. They remain relaxed as long as the horse goes at the gait and rate, and in the direction desired. Hands softly relaxed in this manner are said to be passive. Contact of unvarying intensity is principally maintained by the flexibility and play of the elbow joints, which smoothly open and close in order to follow the oscillation of the horse's head and neck.

The normal feel on the stretched rein increases as the speed at any gait increases. At the extended trot and the fast gallop the feel on the mouth is quite frank. However this feel is never an effect of pulling on the reins. It is a contact and a support that the horse learns to reach for when the legs demand an increase of impulsion. A horse that seeks such a contact is often said to "take a nice feel on the bit."

The reins being adjusted, the hands *resist* when they are fixed in place, and for this reason a resisting hand is often referred to as a *fixed hand*. They *follow* when they follow the movements of the head and neck.

When the rider can keep a softly stretched rein of unvarying intensity, with hands that are gentle and elastic, he may be said to have good hands. When the hand becomes active for any purpose, it is first moved just the distance necessary to a position where it can best act to produce the result desired on the horse's movement. At this time the legs increase their drive forcing the horse to accept additional contact of the mouth with the bit. Then, assuming that the reins are properly stretched and adjusted in length, the half relaxed fingers close and tighten on them. Finally, the hand is fixed in place. This increased resistance of the hand continues until the horse obeys the hand, whereupon the fingers instantly relax and return to the normal feel as a reward for his obedience. Often times they may close again almost instantly if the horse has not completed the movement desired or in case the opposition has not been entirely overcome. Fixing the hand often requires a simultaneous fixing, or immobilizing, of the elbows and shoulder joints.

The intensity of the hand's resistance must be just equal to that offered by the horse, never more.

Among these numerous rein actions, it is necessary to determine those whose simple and definite effects are sufficient to obtain all of the movements which are useful in military equitation.

The reins regulate impulsion. The two reins, acting together, should have the effect of slowing, stopping, or backing the horse. They are called *direct reins*. This action should be produced by fixing the hands and closing the fingers on the adjusted reins; the elbows and hands should move as little as possible to the rear.

The hands control also the position of the forehand.

The reins act through the mouth on the head, neck, and shoulders; they permit the displacement of the head with respect to the neck; the neck with respect to the shoulders; the shoulders with respect to the haunches. They may even act indirectly on the haunches by giving the shoulders such a position that the haunches are obliged to change direction, which is called *"opposing the shoulders to the haunches"*[3]

These different effects depend on the direction of tension on the rein, according to whether the hand is carried more or less forward or to the rear, to the right or left, high or low.

One may group the several actions of the hand into five principal effects, but it is a purely theoretical division which facilitates instruction in the study of the aids. These five effects are capable of an infinite number of variations, depending upon the exact amount and direction of the force applied, as well as upon the various combinations of effects employed. They are further modified by the actions of the rider's legs. Despite the numberless variations thus made possible, it is very useful to have a knowledge of how to produce these five effects, and of their results on the horse. With practice, their use, singly and in combination, ultimately becomes instinctive.

The effects of the *right rein* are used as examples throughout.

---

3   The term "opposition" as used in connection with rein actions implies an effect of opposing the shoulders to the haunches, which, as is stated above, is produced by "giving the shoulders such a position that the haunches are obliged to change direction." This position and result are produced by rein action which changes the direction of the shoulders (forehand) at the same time that it retards them, implying an increased tension on the rein.

1. (1st Rein Effect)
"Opening" or *"Leading Rein."*

The opening rein is used to turn the horse to the right, from the halt or while moving at any gait. It is the rein effect with which a green colt can be most easily taught to change direction. As it acts in a natural simple manner, it is one of the first employed. (See Sketch 1, Fig. 6).

The right hand is carried to the right and slightly to the front. There is no tension or pulling to the rear. It turns the horse's head and bends his neck to the right, throwing the bulk of their weight on the right foreleg. This tends to make him lose his balance to the right, and causes him to move in that direction to regain his equilibrium. If in motion, the horse turns to the right on a large curve, the body following naturally after the head and neck. When using this rein effect, the rider's two legs keep up the horse's movement, usually acting with equal force. However, if a rather sharp turn is desired, the right leg is used more vigorously against the horse's side, and a little farther to the rear. This pushes the hind quarters to the left more rapidly, and makes the curve over which he is moving, sharper. If the horse is at a halt, and the rider does not use his legs, this rein will simply turn the head to the right, and weight the right shoulder.

The left hand, normally passive, should be kept low. (Most riders pull on that rein or carry it across the horse's neck, which counteracts the effect sought with the right rein). If the horse bends his neck too far to the right, the left rein intervenes by being fixed at an appropriate length, low and close to the left shoulder, in order to limit the neck's bend to the right. This is often necessary with a green colt, and prevents his becoming "rubber-necked" (bending the neck too much at the shoulders, instead of uniformly throughout its length).

The effect of the opening rein is to lead the horse to the right front, not to pull him around. A common fault in riding is pulling when changing direction although no decrease in speed is desired.

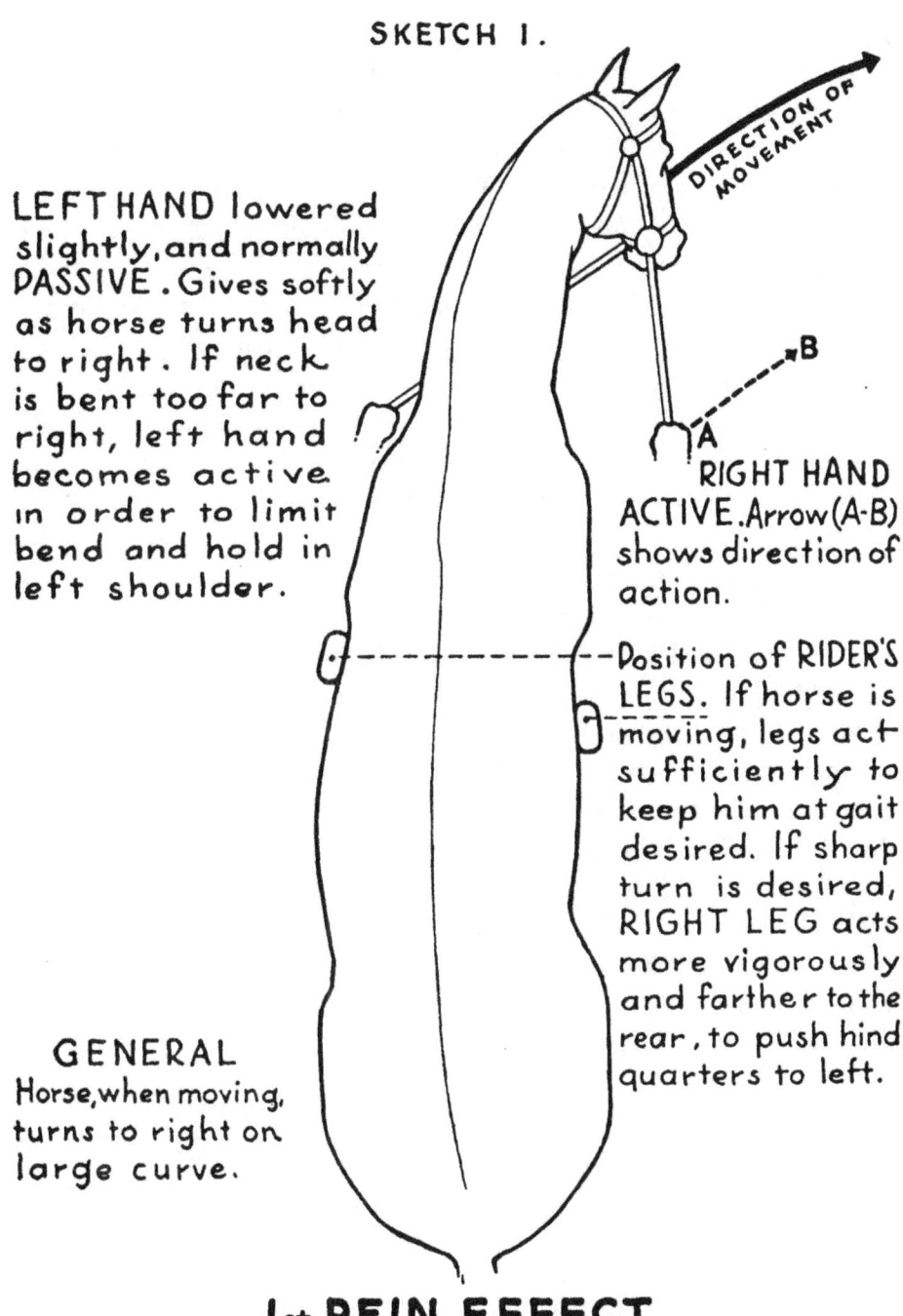

1st REIN EFFECT
Right Opening, or Leading Rein

Fig. 6.

2. (2nd Rein Effect)
"Direct Rein of Opposition"

The right hand, held normally a few inches higher than the withers, is carried slightly to the right, and drawn to the rear, to a point where it becomes effective. The hand is then more or less fixed, thus bringing the horse's nose to the right and rear. The neck, perforce, bends to the right as a result, and the weight of the head and neck is thrown on the right shoulder. This weight impedes the action of the right shoulder and leg, and also breaks the horse's equilibrium to the right. Automatically his hind quarters are forced around to the left. If he is at a halt, the action will turn him about in place; forelegs moving to the right, hind legs to the left.

If moving, the horse is forced to turn to the right. The sharpness of the curve is regulated by the amount of tension on the right rein. The hind quarters being forced out to the left, the turn is made, more or less, "on the forehand."

This is a powerful rein effect, and should be taught all horses, as it is irresistible when the right leg, or spur, if necessary, is used to force the hind quarters to the left. The horse is compelled by these aids to turn very sharply to the right. When, through fear or obstinacy, he resists turning, this rein effect and the spur should be used.

The left hand, held above the left shoulder, gives passively to the front as the head turns to the right. It should not resist the action of the right rein *unless* the horse bends his neck too far to the right, in which case the left rein becomes active, and limits, as necessary, the bend of the neck. (See Sketch 2 Fig. 7). With a horse heavy on the forehand, and inclined to pull, the right hand should be well raised, in using this rein effect, to raise the head and decrease the load on the shoulders.

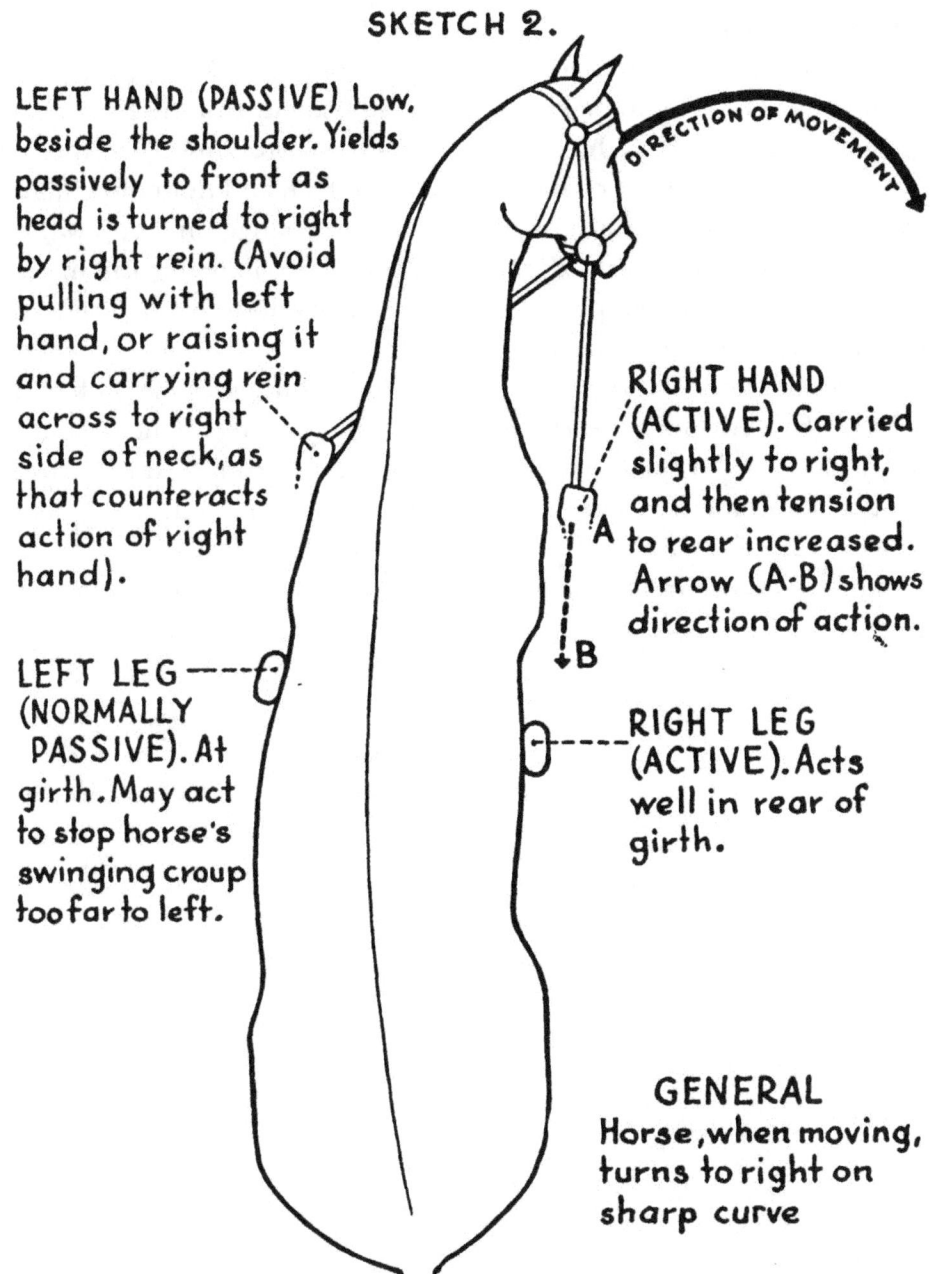

Fig. 7. 2nd Rein Effect.

3. (3rd Rein Effect)
"Bearing" or "Neck Rein"

The right hand is carried just over the crest of the neck, and acts toward the left front. The rein, to be effective, should bear against the right side *of the upper half of the neck,* as this part of the neck is more sensitive to the rein than that near the shoulders. It is an artificial effect, and not powerful, but is the one habitually used with trained horses, to change direction without changing speed, particularly in polo. By using the *left opening rein* in early training, and later combining the *right bearing rein* with it, obedience to the *right bearing rein* alone is easily taught.

It turns the horse's nose upward to the right, and forces the bulk of the weight of the neck onto the *left* shoulder. While its effect is not strong, if moving, the horse's balance is shifted toward the left front, and he turns on a large curve to the left. The rider's legs normally remain in place, acting only to sustain the gait. (See Sketch 3 Fig. 8). The effect should be produced intermittently each time the left foreleg is moved when working with a green colt.

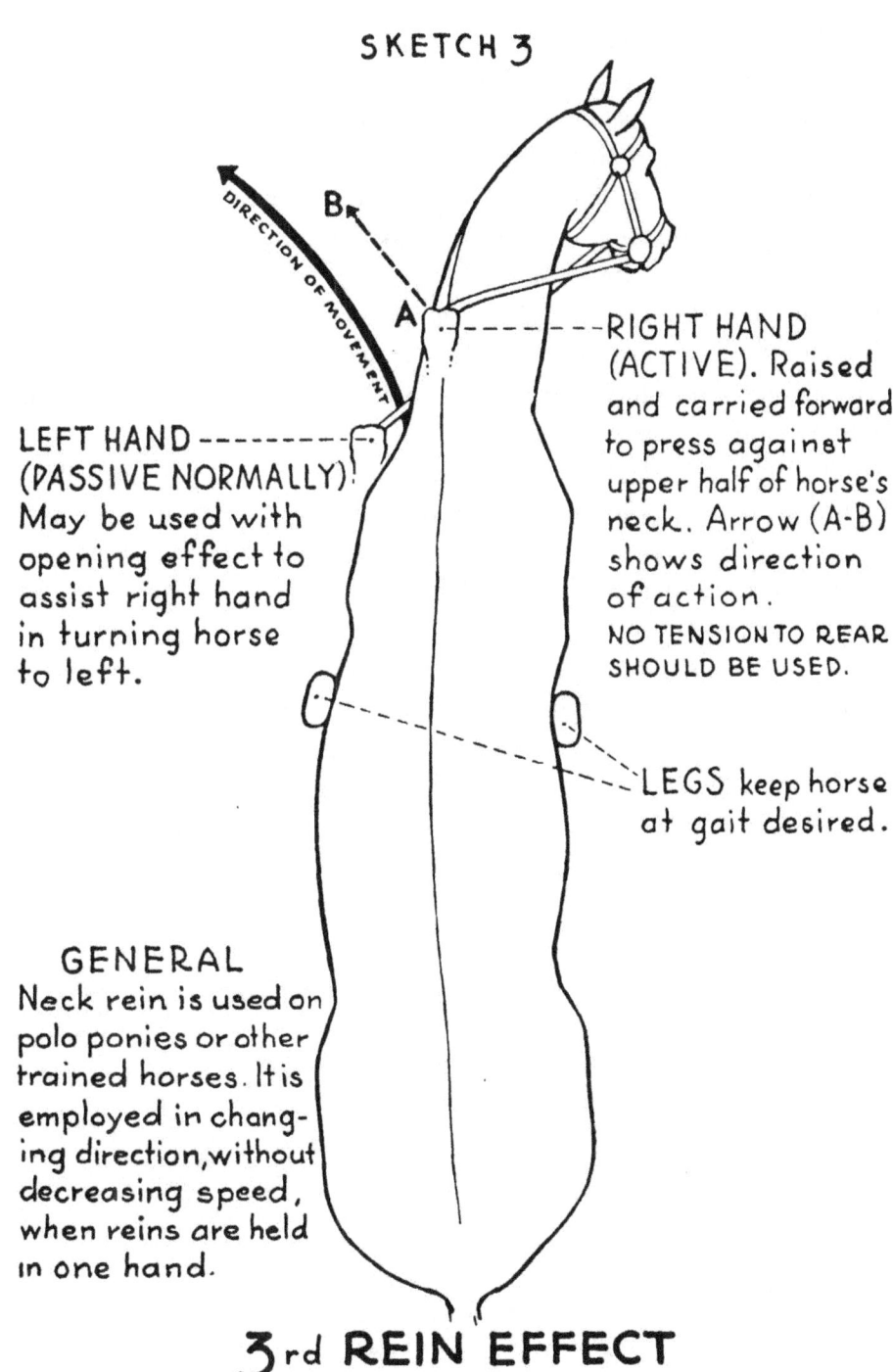

Fig. 8.

4. (4th Rein Effect)
"Rein of Indirect Opposition, in Front of the Withers"

The right hand, slightly raised, is carried slightly across the neck to the left, *in front of the withers*. The fingers are then closed causing the horse's nose to turn to the right and rear.

The mass of the horse's head and neck are forced against the *left* shoulder, impeding its action and tending to break the horse's equilibrium toward the left.

If the horse is standing still, *the rein of opposition in front of the wither** tends to push the shoulders around to the left rear, and the hind quarters, due to the movement of the shoulders, are automatically forced to the right front, the horse turning in place on his center.

If moving, the horse turns to the left, the sharpness of the turn being regulated by the amount of tension on the right rein. In turning sharply to the left, the left rein can also aid by acting to the left rear, and *parallel* to the right rein. As the horse becomes well trained, the left rein should become more and more passive.

The rider can force the horse to turn on his center by using his left leg to push the croup to the right. If, however, he is teaching a "turn on the haunches," the *right* leg acts well in rear of the girth to keep the haunches inside and prevent their swinging out on the turn. This rein effect is used to turn a trained horse when the reins are held in one hand, as in polo. (See Sketch 4 Fig. 9).

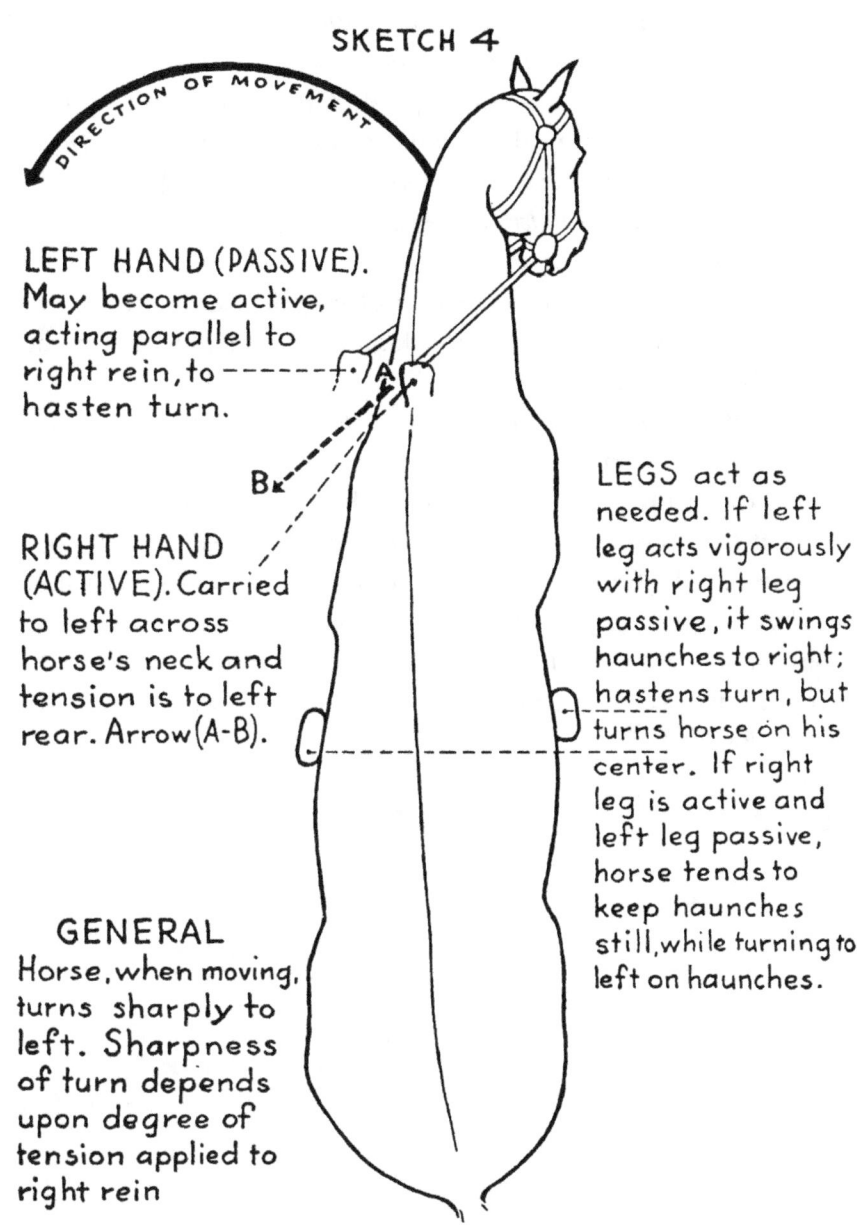

Fig. 9. 4th Rein Effect.

5. (5th Rein Effect)
"Rein of Indirect Opposition, in Rear of the Withers"

The right hand is kept to the right of the withers, although the rein acts obliquely to the rear and left, toward the horse's left hip. The rider's right leg is active and helps drive the horse's hind quarters to the left. *His head turns to the right and rear; the neck and backbone are curved to the right, forcing the mass of the horse against his left hip and hind leg,* while increasing the weight borne by the left shoulder, as well.

The left rein is normally passive, but may act parallel to the right rein, or, more often, as a "leading rein," (1st Rein Effect), to assist in moving the horse to the left front.

The right leg, and right rein of indirect opposition, in rear of the withers, have a powerful, dominating effect. A moving horse, when "bent around the right leg" by their action, is overbalanced to the left, and must move diagonally to the left front crossing his front and hind legs respectively, to retain his equilibrium. It is impossible for a well-trained horse to resist their action, or to shy toward the right when these "aids" are applied. His whole body is bent like a bow, the rein and rider's arm forming the tightened string. When moving, he is forced to chase his own balance toward the left front. (See Sketch 5. Fig. 10).

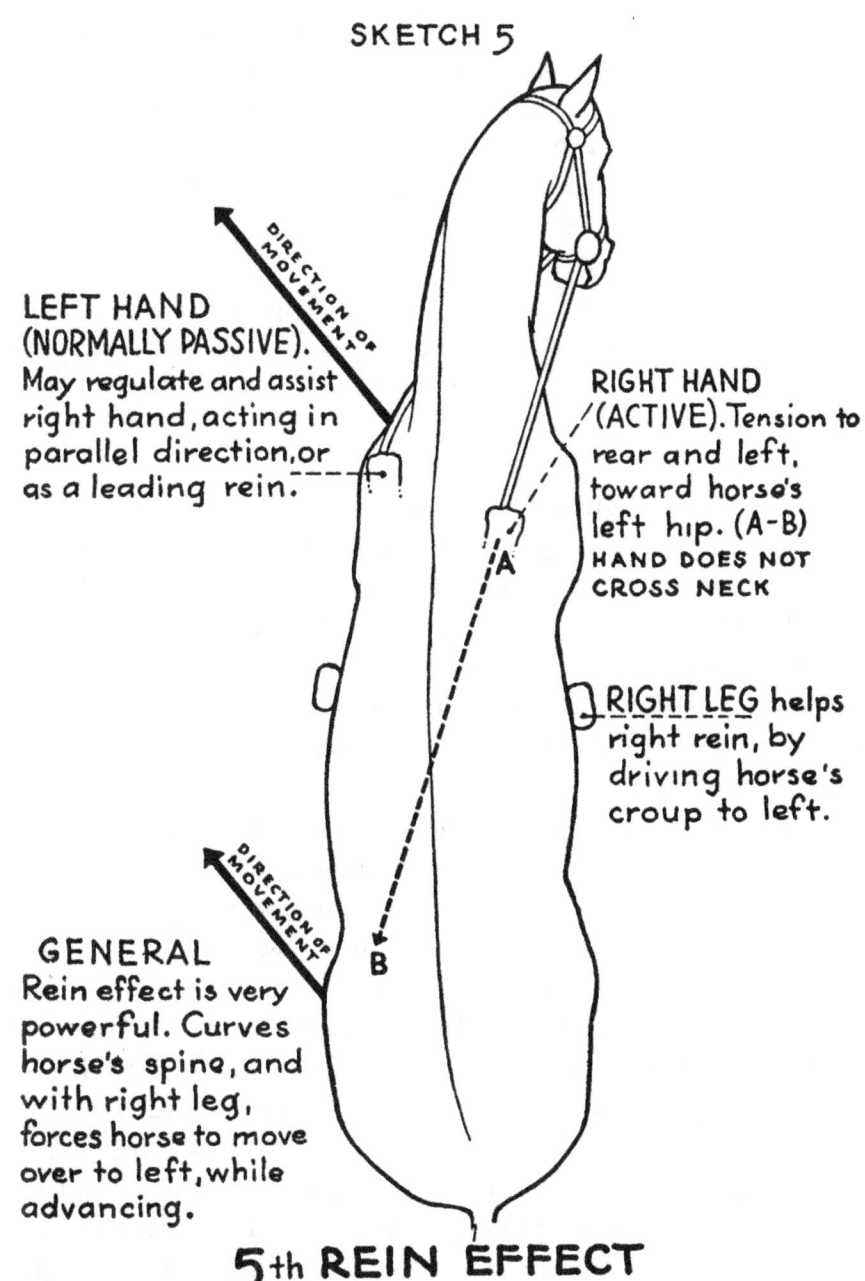

Fig. 10.

These five rein effects can gradually be merged, one into another, and produce an infinite number of different effects. The "passive" rein must come to the assistance of the "active" rein at times, and the rider's legs act to drive the haunches over to one side or the other, depending upon the horse's actions. The reins must be shortened or lengthened, as necessary. For example, a very short hold must be taken on the bearing rein, when training a green horse. The rein also must be shortened to efficaciously apply the "Fifth Effect."

Any horseman with theoretical knowledge of what he wants, can by practice, master the rein effects and soon appreciate the methods of combining them.

c. The weight

In studying the actions of the reins, it has been seen that under their influence the balance of the horse can be modified even so as to lead him to turn to the right or left accordingly, as the weight of the neck is carried on one shoulder or the other. As the shoulders are unequally weighted, the forehand moves towards the side to which the excess of weight draws it.

The equal or unequal distribution of the horse's mass upon the supporting members evidently has a direct influence on the direction of movement taken by the whole machine.

When the horse carries a rider, the mass which the members support does not consist of the horse's weight alone; to that is added the weight of the rider—between 165 and 190 pounds, on the average [male]. The trunk which alone amounts to about 100 pounds may, by shifting its position, contribute materially to the variations in the balance of the horse provoked by the aids. The rider, then, must be warned not to hinder the movements of the horse by an improper distribution of his weight, but, on the contrary, to favor them by using his weight always in the desired direction.

When moving, stopping, turning, or on two tracks, the rider, by slightly displacing his weight in the direction of movement, may facilitate and hasten the obedience of the horse. While quite clearly marked in the breaking and training of a young horse, these displacements of the weight become more and more limited as training is perfected.

In superior equitation, they are reduced to a mere *weighting of a stirrup*.

23. ACCORD OF THE AIDS.

The "accord of the aids" is that cooperation which should exist between the rider's legs, hands, and weight, which will permit, facilitate, or hasten proper execution of the movements desired.

a. Accord of the legs acting together and the two reins acting together.

The legs give impulsion. The reins regulate the impulsion. The action of both legs together has the effect of producing, maintaining, or accelerating the forward movement. Impulsion is forward movement subject to the exact discipline of the aids and exploited in view of the object to be attained.

Tension on both reins together has the effect of limiting the forward movement; that is, of slowing, stopping, or backing.

These two actions (simultaneous action of both legs and both hands), then, are totally opposed and should never be produced at the same time under pain of destroying the impulsion.

When the legs *act* to increase speed, the hands should *yield* to allow the increase; then they *resist* if necessary, to limit it.

Likewise, when the reins *act* to slow the gait, the legs *yield*, then *resist* if necessary, to limit the decrease.

Summing up: in slowing, stopping or backing, the legs oversee the movement in order to regulate it.

When moving forward, taking the trot, or otherwise increasing the gait, the reins should be ready to resist at the proper moment in order to regulate the gait, but they come into play only after the horse has commenced to yield to the action of the legs.

On straight lines, therefore, the actions of the hands and the legs should never be simultaneous.

It is evident that the more obedient and highly trained the horse, the more these actions may approach one another without confusion. The "greener" the horse, the more distinct the indications given him should be and the greater the necessity for separate actions of those aids whose effects might be contradictory.

b. Accord of the two reins.

When seeking to regulate or reinforce the action of one rein by that of the other, care must be taken that they do not contradict each other; if the right hand acts, the left must allow the right to produce its full effect.

Consequently, the left hand not only should not *resist;* it should *yield.* If it acts simultaneously with the right hand, far from strengthening the action, it can only oppose, weaken, or even destroy it.

Conversely, the yielding of the left hand, when the right acts, allows the action of the right hand to have its full effect.

So, whenever the right rein acts, whether as opening rein, bearing rein, or rein of opposition, the left hand should at first yield to permit the head and neck to take the indicated position, then resist, if necessary, to limit the movement. It then plays the role of the regulating rein.

c. Accord of the two legs.

When the right leg acts alone, the left leg should, at first, yield to allow the action of the right leg to produce its effect, then resist, if necessary, to regulate the movement by limiting the displacement of the croup.

d. Faults to avoid.

The rider is cautioned about the faults to be avoided at the beginning, in order not to spoil his horse definitely and permanently, but to teach him exact obedience to the aids when they act with accord.

The rider must:

(1) Never pull on the reins. This has been explained above, as has been the manner of action of the head.

(2) Only make use of the equal action of the legs when impulsion must be obtained or increased, either to put the horse in movement, change the gait, increase the speed, or counteract a slowing up of the horse caused by the direct action of the reins.

(3) Not allow the neck to bend all the way from the shoulders, nor to bend in a pronounced manner, through the preponderant action of the rein. To prevent this, the other rein must assist to obtain the desired result, as will be explained in changes of direction.

(4) Not forget, when using one leg with greater force than the other, that the latter oversees the action produced by the former and is ready to assist when necessary in the maintenance of the impulsion, gait and speed.

24. LATERAL AND DIAGONAL AIDS, LATERAL AND DIAGONAL EFFECTS.

In instruction, to shorten explanations, the aids are considered either from the viewpoint of the various combinations which may result from the association of the two hands and the two legs; or, from the viewpoint of the direction of their action, that is to say, of the effects produced.

When the determining aids are placed on the same side of the horse, right leg and right rein, they are called *lateral aids*.

When they are on the contrary, one on the right, the other on the left of the horse; for example, left leg, right hand, they are called *diagonal aids*.

Considering rein actions alone, with respect to the *direction* of their actions and the *effects* produced; when the *direction* of the hand action is on the same side of the horse as the hand acting (in other words, is parallel to or away from the horse), a *lateral effect* is produced; example—opening rein, direct rein; when the *direction* of the hand action is *towards* the horse, a *diagonal effect* is produced.

Diagonal effect includes all actions of the hand in the direction of the horse; the right hand for example, acting diagonally from front to rear and right to left (actions of the indirect reins of opposition).

Following these definitions, if in the two track towards the right, the rider uses his left leg and left rein, he employs *lateral aids;* but the left hand acting diagonally from the front to rear and from left to right produces a *diagonal* effect.

If, in the same movement, the rider uses his left leg and right rein, he employs *diagonal* aids; but the right rein in leading the head produces, in the direction of march, a *lateral effect*.

True equitation is nothing more than the combination of the different lateral effects or diagonal effects of which we have just been speaking. The rider has two hands and two legs which may act singly or together, laterally or diagonally, and thus produce very varied effects. It is "up to the rider" to use, according to the horse he is riding and the purpose in view, the aid or the aids which should produce the desired effect.

25. ARTIFICIAL AIDS.

The artificial aids are the means of domination created by the industry and ingenuity of man to prolong, strengthen, or take the place of his natural aids. They vary with the nature of the horse and the use made of him.

Those which have a current use are first, the *riding whip*, much used in the beginning of training to teach a young horse to yield the haunches to the action of the leg, as well as an aid in producing forward movement, and in ordinary riding with mares and sensitive horses who kick at the boot; then the *longeing whip, link straps, martingales, nosebands, rigid reins, pulley or running reins, etc.* Included also are the various kinds of spurs, as well as the innumerable types of curbs, gag snaffles, rearing bits, etc.

These different means may be useful to quickly prepare a horse for service, to rapidly reassert lost authority and to dominate certain difficult horses in order that their training may progress. But it must not be forgotten that most of these instruments, excellent in certain hands, become dangerous with the less experienced.

Besides, the results, even though rapidly obtained by these means, are generally only superficial. They cannot really take the place of the true education of the horse, which depends as much upon his moral submission as upon his physical obedience to the natural aids.

CHAPTER VII

THE MANAGEMENT AND CONTROL OF THE HORSE

26. Feeling the horse's mouth 68
    a. Adjustment of the reins 68
    b. Hands: Active, passive, fixed, educated 69
    c. Length of reins with the fixed hand 71
    d. The Flexions 72
        (1) Softening or flexing the lower jaw by
            vibrations 73
        (2) Direct flexion of the poll 74
        (3) Lateral flexion 76
    e. The gather 78
    f. Descent of the neck and placing the horse's head .. 79
    g. The horse collected 81
    h. The half-halt 83
27. Putting the horse in motion and
    increasing the gait and rate 83
    a. Being at the halt, to walk or trot 83
    b. To increase the rate of the walk, the trot, or the gallop .. 84
28. Decreasing the gait and rate 84
    a. To halt from the walk 84
    b. To come to a slower gait from the trot or the
        gallop or to halt 85
    c. To slow the walk, trot, or gallop 85
29. The turn 85
30. Ranging the haunches 87
31. The about on the forehand 88
32. The about on the haunches 89

33. To cadence the trot 91
34. To extend the trot 92
35. To back.. 92
36. The shoulder in 93
37. Work on two tracks 95
38. The haunches out 97
39. The haunches in... 97
40. Gallop departs 97
41. Gallop departs with the horse straight......................... 99
42. The false gallop 100
43. To cadence the gallop 101
44. The change of lead.. 102
45. Obtaining balance... 103
46. Remarks on the gaits.. 104
47. Extracts from the regulations of the F. E. I................. 105
48. Mastery of the aids... 108
49. Equestrian tact.. 110

26. FEELING THE HORSE'S MOUTH.

a. Adjustment of the reins.

The rider should have the reins sufficiently stretched to maintain contact with his horse's mouth at all times. This is an absolute rule, exception to which is made only when the reins are completely abandoned. It is only through the tension of the reins, no matter how slight, that the rider may feel the horse's mouth, which is the best medium for indicating his intentions. Also, the rider is able to act immediately and smoothly on the mouth only by means of the uninterrupted contact between his fingers and the horse's mouth. Moreover, this contact keeps the horse dependent on his rider by causing him to feel the means of the rider's domination which his training has taught him to respect.

Finally, if the hand does not at all times regulate the gait, but allows it to be modulated at will by the horse, his impulsion is lost. Only when the horse's jaw yields, as will be dealt with under flexions, is contact lost with his mouth. This, however, is a momentary loss of contact only.

The necessity of maintaining contact with the mouth does not necessarily mean that the horse should be always ridden with a strong tension on the reins. The reins should not become floating, however, no matter how low the neck and head stretch out. Exception to this rule of contact, as stated, is made only when the horse yields his jaw, is at rest, or is being intentionally allowed to go on a floating rein for relaxation.

The reins should be adjusted in order to obtain contact with the horse's mouth. By this is meant a length of rein, which varies according to the kind of work undertaken, but which permits a soft feel of the horse's mouth.

b. Hands: active, passive, fixed, educated.

When the horse is going in the direction and at the gait and rate desired, the hands are "passive," and "give and take," as the expression goes, so as to follow the horse's mouth with the same even tension of rein. Remaining "passive" under these circumstances, as just indicated, is the principal characteristic of "good hands."

However, when the horse is disobedient, or when an effect is produced on his mouth to demand a decrease of gait, change of direction, or some particular movement, *"educated" hands do not "give," but resist, as long as the horse "takes." They only "give" when the horse "gives," or ceases to "take."*

When the hand becomes active for any purpose, it is first moved *just the distance necessary to a position where it can best act to produce the result desired on the horse's movement. Then, assuming that the reins are properly stretched and adjusted in length, the half-relaxed fingers close and tighten on them. Finally, the hand is "fixed" in place, maintaining the additional resistance thus produced against the mouth with the bit. This increased resistance continues until the horse obeys the hand, whereupon the fingers instantly relax as a reward for his obedience.* Oftentimes, they may close again almost instantly, if the horse has not completed the movement desired, or in case the opposition he has presented has not been entirely subdued. Fixing the hand often requires a simultaneous fixing, or immobilizing, of the elbow and shoulder joints, when the horse's resistance is strong and stubborn. In other words, the whole arm is "set." In this case, the whole arm also relaxes simultaneously with the fingers in rewarding the horse.

The above method of using the hands is very simple, after some practice, and is much more efficacious than pulling with the weight of the body and strength of the arms. In pulling, instead of resisting by fixing the hand, the rider usually does not detect the horse's concession. As a result, his hand, instead of relaxing, flies to the rear when the horse gives. No reward follows the horse's giving; consequently, with pulling hands, the mouth is never improved. A man who pulls, instead of resisting with fixed hands, may be likened to one who is pulling on a rope against an opponent in "tug of war." If the opponent unexpectedly lets go his end of the rope, the puller falls backward. If, however, the latter has prepared for such a contingency by remaining in balance and using his muscles rather than his weight in resisting his opponent, he will not be surprised and fall backward. *This should be the thought in using the reins; to maintain a balanced seat and resist, rather than pull.* The fixing of the hands need not be absolutely rigid and immovable in all cases. Tact must be used; the finger, elbow, and shoulder joints set up a resistance more or less fixed, depending on the type of horse, and his degree of training. If the lightly-fixed hands are pulled slightly out of their position by a young, partly-trained horse, they are promptly replaced and fixed a little more firmly.

The more highly educated the hands become, the more quickly and smoothly they resist or relax at the appropriate time, and the more accurately they measure the correct amount of resistance. *In many cases where the hand has been carried to the rear and fixed, it is necessary to relax the whole arm and move the hand forward in addition to relaxing the fingers, in order to cede entirely to the horse's obedience.*

Where the horse does not at once obey the action of the hand, it continues to *"resist"* by remaining fixed. It steadily maintains the increased tension on the taut rein at exactly the same intensity until the horse yields. A series of pulls and givings never teach a horse anything. Often with the fixed hand, the fingers may "work the bit" a trifle to break up tactfully the resistance of the horse's jaw, as will be discussed under "Vibrations." *The intensity of the hand's resistance must be just equal to that offered by the horse; never more.*

The ability to fix the hand in the necessary place, with a resistance exactly equal to the horse's resistance, and to yield the exact instant the horse yields, is the whole secret of "educated hands." Without this ability, the true art of riding, and the feeling given by a perfect mouth, are unknown.

Both hands, or one at a time, may resist, depending on what is desired. The hand that is not active is normally soft and passive, *giving what rein is necessary* to allow the active hand to place the horse's head in the position suitable for executing the called-for movement. Take, for example, a turn to the right by direct tension to the rear on the right rein: as the horse turns his head to the right, due to the resistance on the right rein, the left hand moves to the front slightly, in order to let the hand turn. This passiveness prevents the left hand's contradicting the right.

A hand may have to move to the left and up; to the right and rear; to the right and forward, etc., so as to act in the correct direction against the bit in securing a certain movement from the horse. *Only after it is correctly placed, do the fingers close on the stretched rein to augment the tension.* Let it be repeated that *only* the resistance necessary to make the horse obey—no more, no less—is used by the *skillful rider.* This may equal only a few ounces on an obedient, well-trained horse, while on a spoiled horse, a hundred pounds may be required. No matter how obstinate the horse may be, the instantaneous yielding to his obedience, whenever it comes, provides the only way of improving his mouth and manners.

"Educated hands" can be acquired by any moderately good horseman, but not in a day or a week, and *only through thinking when in the saddle and reflecting on results obtained after a ride.* Such hands can quickly make a perfectly trained horse out of a green colt, and ultimately reclaim most bad-tempered and spoiled horses. The only necessary adjunct to assist educated hands are *the legs,* which must act to maintain the impulsion at all times.

To recapitulate; the keynote of educated hands is "resistance," not "pulling." To establish only the resistance equal to the force with which the horse is opposing the rider's will, requires a "fixed" hand. The "fixed" hand is placed in the appropriate position to obtain a certain reaction from the horse. There, with just the proper resistance, it remains still, relative to the horse's mouth, and undisturbed by the movements which he makes with his head, neck, or mouth in order to escape the bit's action. As frequently emphasized, the hand can only be fixed when the rider has a secure, steady seat and independence of the aids.

In extreme cases, it may be necessary to fix the hands on the saddle pommel or on the horse's neck to overcome an obstinate or spoiled horse. With such a horse, when the hands are thus fixed, the *heels of the thumbs should be placed against the saddle or neck so that the fingers may feel when the horse yields* and instantly relax to reward him. It is also helpful to clamp the elbows or wrists against the body when first learning to fix the hands. This can be done easily at the walk or slow trot. While the reins are usually a little too long with the hands in this position, it permits the rider to get the "feel" of the "fixed" hand.

c. Length of Rein with the Fixed Hand.

One of the greatest errors committed in using the fixed hand is that of having the reins too short when the resistance is set up. Never-ending effort should be made to avoid arching a horse's neck too much. With a young horse, particularly, great care should be taken when the hands are fixed. They should not be drawn an inch farther to the rear than is absolutely necessary to accomplish the desired result. Oftentimes a young horse, through ignorance or high-spirits, or an older horse, through ignorance or obstinacy, will not immediately obey the hand's fixed resistance when applied with fairly long reins. However, if the rider patiently perseveres, the horse will normally obey after a greater or less period of time. If an old horse is very hard-mouthed, it will be necessary to use a bit and bridoon to compel obedience. In any event, set up your resistance and stick to it; even if in the first lessons, it is a matter of one, two, or three minutes actual time before the horse yields. If you feel you must shorten the rein or reins to make him respond, shorten them only a very small amount at a time, then fix the hand again until he yields by softening his jaw, turning, slowing, or halting, as the case may be.

Above all else, avoid folding his neck and head into a ball against his chest. It is surprising how quickly, with a little patience, a horse can be taught to yield to the bit with long reins while keeping his neck extended.

d. *The Flexions.*
There is scarcely a horseman who has not had many occasions to note the ease with which his horse is managed when the indications of the bit are received with flexibility, nor the opposite result when the poll and the jaws are contracted.

The reason for this is that if the neck and head remain rigid in all their articulations, they form practically a rigid part of the rest of the body. The propelling power of the hind legs is transmitted to the rider's hands without any diminution in force, and they, in a like manner, must react on the hindquarters with great energy in order to control them.

Under these conditions the rider is contending with the driving force of the horse, the full effect of which is transmitted to him in such a way that the horse cannot be guided with any gentleness.

On the other hand, if the articulations of the jaw and the poll are supple, they become an intermediary between the hindquarters and the bit, and by their elasticity not only lessen the thrust of the mass driven onto the hand by the hind legs, but also join its force to that of the fingers, thus allowing the latter to control the hind legs, while remaining light in their action.

This action is somewhat analogous to that which takes place when a railroad car is pushed against a bumper in such a way as to remain in place. The buffers, while being compressed, lessen the shock received from the bumper and, in addition, store up a force which lessens the amount needed to withdraw the car. This is similar to the case of the neck and the jaw, which lessen the drive of the mass onto the hands and increase their control over the hind legs in such a way that the rider's efforts become extremely light. This is the extreme of lightness, and the ultimate purpose of the flexions.

The flexion of the poll and the jaw is called "direct flexion" if it is produced in the vertical plane of the axis of the horse, and "lateral flexion" if it is produced in an oblique plane.

(1) Softening or flexing the lower jaw by "vibrations."

When the fingers are closed on the adjusted reins, a well-trained horse, instead of resisting their action by contracting his jaw, relaxes it. His mouth partly opens, and when the rider's fingers yield in answer to this concession, he gently closes his jaw again, and softly chews once or twice on the bit, as though feeling it, to be sure the rider has loosened the rein. *This softening or flexion of the jaw should always precede the arching, or "direct flexion" of the neck at the poll,* which will be discussed later. It should therefore always be taught the colt before teaching "direct flexion."

"Vibrations" to relax the jaw are executed as follows: the snaffle bit is worked back and forth through the mouth by a soft, sawing action on the reins. This sawing, or vibration, is produced by closing the fingers of the right hand, for example, while the fingers of the left hand simultaneously relax to let the bit slip an inch or so through the horse's mouth toward the right. Then smoothly, slowly, and without delay, the fingers of the right hand partially relax while those of the left close and slide the bit back toward the left. The wrists may assist the action of the fingers by bending inward and upward so that the palm of the hand turns toward the rider's chest. Vibrations are usually executed with a snaffle bit, and the reins remain lightly stretched, the hands carefully avoiding any abrupt or harsh effect to the mouth. Allowing much slack to exist in the rein creates a jerky, irritating effect which causes immediate stiffening rather than de-contraction of the jaw.

With a little patience, this gentle sawing effected by the fingers on the stretched reins will cause a horse to relax his lower jaw muscles and open his mouth. It is undesirable to force the mouth wide open; it should open only part way. A rider not accustomed to this work and whose hands are not trained to feel when the jaw yields, should watch the horse's mouth in order to see when it opens. The opening should be *instantly* rewarded by ceasing the vibrations and a momentary decrease in the tension on the reins. The cardinal principle of instantaneous reward for obedience applies. At first, this exercise should be executed while the pupil is at a walk or slow trot. For two reasons, it will expedite the softening of the jaw in many cases to perform the vibrations while moving on a circle of about ten yards radius. First, it facilitates seeing the horse's mouth since his head is turned slightly in on the curve over which he is traveling. Secondly, in order to keep the horse on the circle, the action of the rider's inside hand will dominate. Where one hand acts more strongly than the other it favors a quicker response to the vibrations than when the two hands act evenly, as when moving on a straight line. After rewarding the horse for softening the jaw, the exercise should be repeated at short intervals of time. Very soon *it will only be necessary to close the fingers on both reins, without vibrations, to secure a jaw flexion.* No attempt should be made to arch the horse's poll at this time. On the contrary, he should be encouraged to keep his neck and head well to the front and rather low in a natural, unconstrained position. Later, the jaw flexion should be practiced at the normal trot and canter. With some horses, results can be obtained more quickly at the slow trot than at the walk.

As a general rule, the vibrations are executed without any increase over the normal feel in the tension on the reins. However, in the case of older, poorly-trained, or stubborn horses, it may be necessary to increase the tension when vibrating the bit in order to compel a yielding of the jaw. In these cases, the rider's legs should urge the horse along in order to prevent his slowing the gait or halting as a result of the increased tension on the reins.

After a horse has learned to promptly relax his jaw in response to the vibrating rein, if it is desired to slow his speed or to halt, the vibration effect should be given first, then the fingers should be closed and fixed to indicate that a decrease in gait is desired. Normal feel should immediately be established as soon as the horse decreases his gait. The jaw flexion is taught as a preliminary step to decreasing the gait or halting and is the second step in producing a good mouth. The first step is teaching the colt to accept the normal feel on the bit, with an extended neck. The decrease in gait should only be demanded after the jaw has relaxed and given to the action of the hand.

Thus it is seen that vibrations teach a soft, relaxed yielding while obeying the rider's hand. This replaces the horse's natural, instinctive stiffening of the jaw against the bit's action.

The yielding eliminates heavy pulling and the rough action of the bit which is unavoidable with stiff-jawed horses. *As stated, after having perfected the lesson of relaxing the jaw in answer to vibrations, it will be found that soon the same result is obtained by simply increasing the tension on the reins.* Vibrations are a means to an end, and, with the trained horse, are only resorted to from time to time when it is necessary to break up an occasional stubborn or whimsical stiffening of the jaw to which any horse inevitably will return.

(2) *Direct Flexion.*—Direct flexion is the yielding of the poll and the jaw in the vertical plane of the axis of the horse, and may be called the third

step in developing a good mouth; the first being to teach the horse to take a normal feel on the bit and the second, to flex his jaw. *It is highly essential to delay direct flexion until the horse has been perfectly trained in the first two steps just mentioned.*

Direct flexion results from fixing the hands and causing the horse to arch the upper third of the neck in the general region of the poll. The lower two thirds, and particularly the portion just in front of the shoulders, should remain firm and stiff. This lower part should *never* be arched because that lowers and places the head close to the breast. This, in turn, makes him heavy in front and poorly balanced.

When a horse executes direct flexion correctly, his mouth first opens part way as the jaw relaxes, and his head then moves backward, partly closing the angle formed at the throat by the neck and lower jaw. After the head is placed as a result of the flexion, the mouth closes at once. It only reopens when more rein tension is applied. The neck bends only at the poll. The face approaches the vertical but should never remain in rear of the vertical as this condition invariably results from the neck's being over-flexed, particularly near the shoulders. Over-flexion leads to many vices such as pulling, nervousness, falling back of the bit, heaviness of the forehead, and shortened gaits.

The horse must first be gathered in order to obtain the direct flexion. If he is well in hand and given sufficient impulsion, this flexion is obtained by a closing of the fingers. If he does not yield the poll and the jaw, the legs must act to force him to take a stronger contact on the bit which causes him to yield the jaw and the poll in order to escape the resulting pain. When a flexion is obtained by closing the fingers on the adjusted reins, the fingers should be immediately relaxed.

The rider can tell very readily if the flexion is produced, because, during the extremely brief time that the horse abandons and retakes the bit, he has the sensation of no longer having anything in his hand.

The important points in teaching direct flexion are:
1. To first flex the jaw.
2. To fix the hands, with as long a rein as possible to avoid over-flexing the neck. The hands must be fairly high when fixed because usually the bending of the neck will occur above the reins.
3. To maintain fixed hands until the horse is forced to nod slightly by flexing the poll which is instantly and lavishly rewarded by relaxing the hands and by pats on the neck.
4. To use the legs, squeezing with the calves or touching with blunt spurs, if necessary, to keep the horse up to his gait and speed.

Otherwise, he will undoubtedly slow down or stop as a result of the increased tension of the fixed hand. Tact is necessary so that the action of the legs is sufficient to maintain the gait, and so that the resistance of the bit is not so severe as to over-flex the neck or to provoke rebellious defenses.

5. To repeat lessons, progressing very slowly, until the complete direct flexion, as described above, is obtained. This work usually requires several weeks. It should be done at the walk or slow trot and later at the faster gaits; never at the halt with a green colt. The slow trot is probably the best gait as the forward motion can be easily sustained; whereas at the walk, the horse can readily stop if he resists the bit.

(3) *Lateral Flexion.*

Lateral flexion is the yielding of the poll and jaw in turning the head to the right or left when an unequal action of the two reins prevents an extension of the neck.

The jaw yields in lateral flexion as in direct flexion. The poll yields by causing the head to turn one quarter to the right or to the left.

Lateral flexion, as direct flexion, should only be demanded when the horse is gathered and, at first, only at the walk or slow trot. To obtain it to the right, for example, tension should be placed on the right direct rein of opposition which places the head in the desired position; the left rein then acts to limit this movement and to cooperate with the right rein in obtaining the flexion of the jaw.

During the flexion, the right rein is a direct rein of opposition, the left an indirect rein of opposition, and the head is carried to the right of the vertical axis of the horse. This results in carrying the weight of the forehand to the right and in causing the horse to begin a turn to the right.

This turn, during the flexion, can be prevented by having the right rein act diagonally from right to left instead of parallel to the vertical axis of the horse. The weight of the forehand can thus be equally distributed on the two shoulders, leaving the horse free to move straight to the front.

The difficulties encountered are of different sorts. Most often the horse opposes resistance in his mouth and neck, and increases it by his weight by throwing out his left shoulder if, for example, the right lateral flexion is demanded. In this case the legs should be used energetically to give vigorous action, not only to the right rein, but also to the indirect rein of opposition. As soon as the weight is carried to the right, the resistance is broken and the horse yields more readily.

At other times the horse may curve his entire neck from the poll to the shoulder under the action of the aids employed. This is an incorrect flexion and must be avoided. To prevent it, the left reins and the right curb rein should be held in the left hand with the right curb rein held against the neck to prevent it curving to the right. The right snaffle rein is held in the right hand and acts as a direct rein of opposition.

The horse eventually bends his neck less and less at the base and more at the poll, if the bend caused by the direct rein is opposed progressively by the indirect rein of opposition. When this is the case, all that is necessary to obtain the flexion, by means of the regular aids, is to gradually diminish the action of the right bearing rein, so that finally the correct flexion at the poll is obtained by the right direct rein of opposition alone.

When this result is obtained, the action of the left reins assist in obtaining the flexion of the jaw.

It happens frequently that the nose alone cedes to the action of the direct reins, the poll and adjacent parts of the neck remaining in the vertical plane of the horse's axis. This occurs when the poll has not been sufficiently flexed. Its flexion is obtained by raising the left reins so as to cause them to act in opposition against the upper part of the neck.

Finally, at times, the horse seeks to resist flexing to the right by lowering his neck to the left. This is remedied by applying the left indirect rein of opposition at the point where the neck should remain straight.

The lateral flexion is useful for several reasons:

(1) It displaces the weight of the forehand to the side towards which the horse marches. The displacement is caused not only by the action of the direct reins, and that of the indirect reins, but by the bend of the upper part of the neck placing the head outside of the vertical plane of the horse's axis. It is a necessary consideration for oblique or parallel displacements.

(2) The lateral flexion causes the horse to look toward the side to which he is moving. If he did not do this, he would be as difficult to guide as if he were blind. The lateral flexion causes the horse to look toward the side to which he displaces the weight of the forehand, viz:—the direction in which he is moving. Under these circumstances he is able to gauge his stride and regulate his movements.

(3) The lateral flexion establishes between the forehand and the hindquarters a certain degree of independence which, in changes of direction and in two tracking, allows the shoulders and haunches to be given separate movements while closely united to each other. (This is similar to the case of

a whip which is held at the ends and bent in the middle. Both are acting in different directions and yet feel the force acting on each other).

(4) Finally, the lateral flexions accustom the horse to localize in the poll, and parts immediately adjacent thereto, the lateral displacements of the neck.

When the horse has learned this lesson, the neck is united to the shoulders, and when a lateral displacement of the head is desired, the lateral flexions are not made at the withers, which would isolate the neck from the rest of the body and deprive it of its powers of direction.

e. The gather.

The rider must be able to de-contract his horse, at least to the degree necessary to maintain him light and sensitive, in order to place him in proper balance and to overcome resistances which not only make him disagreeable to ride, but allow him momentarily or completely to refuse obedience. In order to avoid these difficulties, the rider must be able to obtain the gather[4], the descent of the neck, and the-flexions, all of which are necessary to permit the horse to be maintained light and sensitive to the action of the aids.

The gather is the result of the action of the fingers and of the legs, by means of which the neck is raised in order to cause the weight to be displaced in the direction of the haunches, or in other words, to obtain the engagement of the hindquarters.

The reins being adjusted, this effect is obtained by substituting for the soft action of the fingers a certain degree of resistance. If the horse reacts to this rein effect by a flexion, a drawing back of the hand, accompanying the momentary relaxation of the jaws forces him to raise the neck in order to again close his mouth. This is the beginning of the gather. Several consecutive elevations, executed in the same manner, increase the gather to a marked degree.

If the horse does not relax his jaw when the fingers are closed, the stretching out of the neck, caused by the action of the legs, is either changed into an elevation which is nothing else than the simple gather without flexion or the neck is lowered and brought in to the chest. This latter case is rare and is corrected by using the snaffle in alternate upward actions on both reins.

Should the horse not respond to the action of the closed fingers, the legs must act in such a manner as to force him on to the bit so strongly that it

---

4   "the gather" is the French "*Rassembler*" concept. The reader is referred to expert writings on the *Rassembler* by Dom Diogo de Branganca: *Dressage in the French Tradition*, Xenophon Press, 2010-Editor's note.

causes him pain and makes him seek to escape it which he does in the manner indicated. The hands may be slightly raised, if necessary, to accentuate their upward action.

The hindquarters are engaged due to the shifting of the center of gravity caused by the elevations of the horse's neck.

It is important that the shortening of the reins, by which the gather is obtained, follows and does not precede the neck in its successive elevations; otherwise there would be traction on the reins with all its disadvantages.

The rider should commence demanding the gather *at the walk* and not at the halt in order to avoid the risk of causing his horse to get behind the bit. The movement of the neck, peculiar to this gait, facilitates moreover the action of the hands.

In the gather, the face of the horse should never come back beyond the vertical, and the poll should always be the most elevated point of the head and neck.

The gather, as it is understood in higher equitation, is not concerned with the direction of the head alone. It concerns itself as much in the submission of the jaw which is the first articulation (or spring) to receive the effect of the hand. If this spring responds with softness to the action which solicits its play, it will bring about the flexibility of the neck and will provoke the tying to it of all the other parts because of the relationship existing among all the muscles. If on the contrary, the jaw resists and refuses to be mobile, then there will be no more lightness, for by nature the resistances mutually sustain themselves, and one resistance will set up numerous others. Thus in higher equitation, the "Gather" represents a general state of submission of all parts of the horse rather than a fixed direction of the head.

The gather commences and is a part of a high degree of collection for which the engagement of the hindquarters is necessary.

f. Descent of the Neck and Lowering the Horse's Head.

The rider must not only be able to displace the horse's weight towards the haunches, but also to displace it forward which is done by the more or less marked lowering of the neck. It consists of an extension and lowering of the neck, obtained by a yielding of the fingers proportional to the extent of movement desired; and the neck being lowered, the weight is displaced towards the shoulders.

A well-trained horse gradually acquires a permanent, graceful, and very slight flexion at the poll when on a stretched rein, while the remainder of the neck is naturally and fully extended. With this head carriage, he is referred to as having his head "well-placed." It is in a position where the bits may act from an advantageous angle, and the horse can easily see everything in his foreground. Due to the relaxation of his poll, back, and loin muscles, he carries himself lightly and with a minimum of fatigue.

Whenever the horse places his head in a position other than the correct one, the hands are moved where they can increase tension on the bit and make his mouth uncomfortable. In these cases, they must be so placed that the horse cannot possibly escape the effect of the bit for a fraction of a second until the rider permits it. When he ultimately seeks to avoid discomfort by putting his head in the correct position—*and then only*—the hands must soften immediately and resume their light, normal feel. In the first instances, it is better to let the reins go slack when rewarding the horse's endeavor to carry his head correctly. This insures his associating the ideas of comfort and correct head carriage.

Here again, the rider's degree of success will be determined by the speed and skill with which he detects the horse's efforts to shift his head to the proper place, and permits it by softening the hands.

To cite one example in applying the rule just given, relative to the action of the hands, take the case of a "star-gazer," (a horse which pokes his nose, head, and neck, high and stiffly in the air, the neck often being "upside down"). Most riders attempt to lower the head by carrying their hands low beside the horse's neck and futilely trying to pull the head down. Nothing could be more foolish. The horse, by tipping the head a little farther to the rear or tossing it suddenly in any direction, can momentarily escape the tension of the reins, ("escape the bit"). Following the old rule of reward and punishment, he will, of course, continue throwing his head as long as he succeeds in escaping the annoyance of the bit even though it may be only for a moment. In other words, he is being taught by the momentary reward he receives that his procedure is correct.

The correct and logical way to lower the head of such a horse is to hold the reins short enough, *(and no shorter),* so that it is impossible for him, by any means, to escape the bit for a single moment. The hands, instead of being lowered in an attempt to pull the horse's head down, are raised, so that, as usual, the forearm and rein make a straight line. *The tension on the rein must become greater than in the normal feel. The hands are more or less fixed and vibrations may be simultaneously employed, all of which increase the horse's discomfort.* The legs compel him to continue at the gait at which he is moving while the hands steadily hold the head in its elevated position. Sooner or later, he becomes tired and uncomfortable in this strained position. Also, he soon discovers that the usual throwing about of the head permits no escape from the bit, and he begins to search for a new way. Finally, he will endeavor to lower his head to a more comfortable and natural position. *Instantly, the hand softens to permit the lowering.*

The horse will doubtlessly make many efforts to raise his head again to its "star-gazing" attitude, but eventually will discover the futility of his efforts to escape the bit which only acts more strongly when his head is elevated. Thereafter, he will maintain the proper head carriage which gives comfort with a less disagreeable tension on the reins. With practice, a horseman can lower, by this means, any "star-gazer's" head in a few minutes. It will, however, often take a long time to redeem a bad case.

As usual, begin the work at the walk or slow trot, then later at the faster gaits. Remember that the horse's impulsion and momentum are low at the walk; therefore, a very slow trot that is easy to sit is often better when trying to place the head or improve the mouth. The hands are more effective when the rider sits quietly and firmly in place. Also, the horse's head remains still at the trot which simplifies the work of the hands. Always have the hands placed so that the effect of the snaffle bit is against the corners of the mouth. To do this, they should remain fairly high at all times when combating a head carriage that is too high. They are instantly lowered to their normal position, of course, when the horse lowers his head.

With an inexperienced rider, the hands and arms may not be sufficiently quick and skillful to relax and follow the star-gazer's head downward when he first attempts to lower it. In this case, *quickly opening the fingers* and allowing the reins to slip through them will insure an instantaneous reward which is vital for success. The reins can be promptly readjusted after the head goes down.

g. *The Horse Collected.*

A collected horse places all his impulsive forces at the disposal of his rider. He is in a state of equilibrium the stability of which may be broken at the slightest demand of the rider by all the forces of the horse which are alert and ready to act.

In this state of collection, the horse is like a whip acted upon by two forces which tend to bring the ends towards each other. If one of these forces ceases, or diminishes in intensity, the whip stretches at once in that direction. The horse reacts similarly. He has an elasticity which is the resultant of all his forces held together by the aids; and if the intensity of one of the aids is increased or diminished, all the live forces concentrated by the state of collection escape towards the side where there is the least resistance, resulting in a change in the equilibrium and sense of movement.

This state of collection naturally requires complete suppleness in the horse and the engagement of the hindquarters. Suppleness, in order to make immediate changes in the equilibrium possible. Engagement of the hindquarters in order to give them control of the mass and the potential to propel it according to the displaced center of gravity.

It is sufficient to state that there is collection only when the neck is raised, and the poll and jaw are completely relaxed. To obtain this, the rider should gather his horse and have him ready to flex his poll or jaw upon the request of the fingers.

Impulsion is necessary in order that the reins act lightly on the mass. In effect, the impulsion holds the forces ready to act in the amount authorized and in the manner desired by the hand. If the contact with the mouth remains light, and if in taking it the horse requires only an indication, the impulsion permits the lightest actions of the hands to provoke a displacement of the center of gravity, just as a feeble weight can oscillate a precise balance, complete balance then is obtained and conserves or changes itself by insignificant effects.

If, on the contrary, impulsion does not exist at all or is insufficient, balance becomes impossible to obtain with lightness; the bit to act must come to the rear; it finds forces opposed to its action. The horse will respond only if the actions of the hands are severe and will obey reluctantly. The rearward action of the hand working under these conditions throws the hocks to the rear. Engagement of the hind legs, as well as balance, becomes impossible except by severe action. One sees from this that impulsion is one of the qualities most necessary for the finesse of equitation and the lightness of the aids.

A horse *should not be maintained in this state of collection for extended periods.* It demands a muscular tension that should cease when the horse is at rest. It should not be employed when the maximum speed of any gait is required, as it requires a position of the neck and a suppleness of the poll and jaw that is unfavorable for speed.

When the rider understands how to obtain and maintain a horse in this state of collection, he is able to displace the weight towards the haunches *by the gather,* towards the shoulders *by lowering the horse's neck,* or laterally by *means of lateral flexions* and the predominant action of one leg—all without effort on his part and, due to the flexions, without resistance from the horse. The aids thus control the horse in the desired manner since they place him in the degree of equilibrium desired.

h. The half-halt.

The half-halt is a firm upward action on stretched reins, with the fingers well closed, followed quickly by a progressive opening of the fingers.

This action is somewhat analogous to those actions one performs in moving a heavy box up a polished staircase when it is essential that the polish of the staircase not be marred nor any noise made. In this case one grasps the box with both hands and lifts up steadily and forcefully until the box is slightly above the next step, one then lowers it steadily and smoothly until it comes to rest without jarring it.

It is used to raise the head and neck and thus carry to the rear the excess of weight that some badly-balanced horses allow to collect on the shoulders. In practice its effect is to improve the horse's balance and to slow down horses that are too ambitious and bearing down into the hand.

It is effected according to need, on one rein, on two together, on the snaffle, or on the curb.

The action of the hand should be regulated according to the resistance which it encounters.

27. PUTTING THE HORSE IN MOTION AND INCREASING THE GAIT AND RATE.
a. Being at the halt, to walk or to trot.

To take up the walk from the halt, the pressure of the legs is progressively increased until the horse moves forward, the fingers being relaxed to allow the neck to extend, but maintaining sufficient tension to keep the horse balanced as he moves out. The horse must be made to move out straight to the front smoothly and without hesitation.

The proper amount of accord must exist between the legs and the reins so that the displacement of the center of gravity and the movement produced by the legs may be regulated and maintained by them to the extent needed to obtain the walk, trot or gallop, as desired by the rider.

The manner of employing the legs must be regulated to the sensibility of the horse in order to execute movement smoothly. By progressive, but rapid, action of the legs to the intensity needed and by yielding the fingers at the moment the neck is extended, a steady forward movement is obtained.

When the horse, with the legs acting equally, displaces his haunches laterally, he is straightened by more intensive action on the part of the leg on the side to which the displacement is made. If the same fault is repeated and if the horse hesitates to move forward, the legs should be used more energetically until the horse executes the proper movement.

The trot is taken from the halt by the same methods.

Careful attention should be paid to the manner in which the horse is put in movement, for if the legs are used with greater energy than is necessary, the movement is executed abruptly instead of smoothly. In such a case the hands must intervene to an amount equal to the excess demanded by the legs. Under these conditions the horse in a short time yields only in a limited way to the action of the legs and loses the valuable asset of obedience to the highest pressure of the legs. In short he becomes, as it is called, *"dead to the legs."*

b. To increase the rate of the walk, the trot, or the gallop.
The speed of the gaits is increased by continuing to relax the fingers when the horse has sufficient impulsion, or if necessary, obtain the needed impulsion by the use of the legs accompanied by the relaxing of the fingers.

This action of the hands must allow the horse to extend his neck as far as is necessary, but the reins must not become floating.

When the desired extension of the neck has been obtained, the horse should receive a firm support from the hand which at the same time follows the movements of the head. In obtaining this support the hands should never be brought to the rear. *The legs must drive the horse up to accept this tension.*

28. DECREASING THE GAIT AND RATE.
a. To halt from the walk.
To execute the halt from the walk, the action of the hand must be changed from lightness to fixity and resistance of the fingers. If the horse halts, the desired result is obtained. If, however, he is but slightly trained, he generally raises his neck in slowing up. The rider then draws his hands to the rear or shortens his reins in order to maintain contact with his mouth and again fixes his hand until the horse halts. *Care should be taken that the rearward action of the hands does not at any time precede the retrograde movement of the mouth,* as in such a case there would be traction on the reins, which, as has been shown, must never be the case. The rider's hands are only drawn to the rear when contact with the horse's mouth is lost, and he desires to reestablish this contact. As soon as the horse halts, the fingers relax their tension on the reins.

At the start of instruction, the rider and instructor need not greatly concern themselves with the exact execution of the halt. Later, however, when progress has been made and it is desired to correct this movement, the rider should pay particular attention to the action of his legs which may not have to act, but which should be prepared to do so if necessary in order to:

(a) insure sufficient contact between the bit and the mouth so that the fixing of the hands, followed by their yielding, will produce either the halt or the elevation of the horse's neck;

(b) prevent an abrupt halt and to insure a smoothly executed movement by decreasing their action as the gait decreases;

(c) act in time and without abruptness should the haunches be displaced laterally, or if the horse tends to back after having been halted. In the former case the leg on the side to which the haunches are displaced acts until the horse is straightened; in the latter case, the simultaneous action of both legs is necessary.

b. To come to a slower gait from the trot or the gallop, or to halt.
These various movements are executed in the same manner as in halting from the walk. However, if a decrease only in the gait is desired instead of the halt, the aids should decrease their action at the moment that the desired decrease in gait is obtained.

c. To slow the walk, trot or gallop.
Decreases in the rate are obtained by the same methods. However, although the horse covers less terrain in the same time, his impulsion must be maintained as otherwise the rider is unable to control the decrease in the desired manner. If the horse commences to lose his impulsion, the rider must maintain it by the proper use of his legs.

When the rider has acquired a certain degree of skill, exercises in decreasing the rate are very useful as they present occasions for determining the exact accord of the aids and of their effects.

29. THE TURN.

The turn comprises two distinct, but closely-united, operations: the change of direction and the march.

It is assumed that the reins are adjusted at the moment the turn is to be executed as they should be at the commencement of any movement. This being the case, the rider has several means of displacing the fore-hand to determine a change of direction. The simplest means by which the beginner can cause his horse to make required changes of direction consists in employing the opening rein by carrying the right hand to the right if a turn in that direction is desired. This draws the head to the right and the shoulders and the rest of the body follow, provided the horse does not resist. The change of direction is thus executed as well as can be expected. One of the disadvantages of this method is that it may displace the haunches to the outside, which cannot have other than unfavorable results.

In addition, the turn by means of the opening rein has the inherent defect of this rein, which is to make it possible for the horse to yield his head and neck and not follow them with the shoulders. The rein aids thus are valueless and the rider no longer has control of the horse.

The turn to the right may also be accomplished by the employment of a left bearing rein. This rein, when used alone, has the same inherent defects as the opening rein in that it is possible for the horse to yield its head and neck and not follow this direction with its shoulders.

Another method employed for a turn to the right is the use of the left indirect rein of opposition in front of the withers. It produces the change of direction and is of great assistance when riding with the reins in one hand. In acting alone, however, it weights the right shoulder more or less strongly, opposes this shoulder to the right haunch, thus slowing up the horse, and necessitating a more active use of the legs than is required when the turn is made in another manner.

The third and best method employed in obtaining the change of direction to the right consists of the combined use of the right direct rein of opposition and the left indirect rein of opposition in front of the withers. The former displaces the head slightly to the right, while the latter prevents this from becoming exaggerated, ties the shoulders to the neck and causes them to take the same direction.

While these actions are taking place, what should be the action of the legs? It has been stated that the turn comprises the change of direction and the march, and that the hands assure the change of direction. The legs must maintain the march. If the action of the reins tends to slow the horse, the change of direction is either executed poorly or not at all. A similar case is that of a boat, not in motion, which does not turn when the rudder is moved to one side or the other. The horse does likewise if he does not advance when the reins demand the turn. The legs should therefore be ready to act, when needed, either to cause the horse to obey the rein aids, or to maintain the gait if it should tend to decrease as a result of the action of the hand. This necessity of advancing moreover makes it obligatory that the haunches deviate neither to the right nor left, or the propulsion of the hind legs will lose their effect. Thus the legs must maintain the hindquarters in their proper place and inclose them in order to prevent, when necessary, their lateral displacement.

If the rider, for example, after having turned to the right, wishes to resume the march to the front, he should distribute the weight of the forehand equally on the two shoulders by the equal and direct action of the two reins. Throughout this action the legs act with equal force, if necessary, to straighten the horse and to push him forward in his new equilibrium.

30. RANGING THE HAUNCHES.

Up to the present time stress has been laid on the action of the legs in producing impulsion in the horse, which they obtain by their simultaneous action, as it is one of the most important factors in riding. However, it is not the only one. In the case where the legs act with unequal pressure, they displace the haunches to the side opposite to the preponderant leg. It is essential that the rider appreciates fully this effect which he employs constantly to move the hindquarters laterally or to displace the weight onto one hind leg, which is of frequent application in equitation.

The rider, first at the walk and then at the trot, practices using one leg with greater force than the other. While moving on a straight line with reins adjusted and legs closed on the horse, he increases the pressure of one leg. This causes the horse to move his haunches to the opposite side proportionally to the degree of increased pressure.

In the beginning a sufficient displacement of the haunches is obtained when the outside hind leg tracks the inside foreleg. For example, if the right leg is acting with greater force, sufficient displacement is obtained if the right hind foot tracks the left fore foot.

The inside leg, the left in the example cited, acts to prevent an exaggerated displacement of the haunches, and also to prevent any slowing up of the gait, and causes the horse to move forward while ranging his haunches.

The necessity of compelling the horse to advance during this movement and of preventing him from placing his shoulders in front of his haunches, is a difficulty which increases according to the oblique position. This is the reason the rider at first is content with a slight displacement.

The role of the reins is to prevent any increase of the gait, while at the same time not hindering the forward movement. It is therefore of little importance, at the start, if the horse moves off the track on which he was formerly moving; the important feature lies in the displacement of the haunches while moving forward.

When a slight displacement of the haunches has been obtained satisfactorily, the amount is gradually increased with out however allowing the angle between the horse's axis and his direction of march to exceed 45°.

Care should be taken by the rider not to use his legs with greater force than necessary. When the outside leg acts with too much force, too great a displacement of the haunches is obtained, and this must be corrected by the action of the inside leg. This produces confusion in the horse, and he soon ceases to range his haunches with the same willingness and does not respond to the action of the legs.

After the rider, through repetition, has learned the effects produced by the lateral action of his legs, he should, in order to confirm himself in the use of his aids acting separately on the forehand and on the hindquarters, practice exercises, such as the abouts on the forehand and haunches while in motion.

31. THE ABOUT ON THE FOREHAND (PIROUETTE RENVERSEE).

The about on the forehand consists of causing the haunches to describe a circle, or an arc thereof, around one shoulder as a pivot.

In this movement the hindquarters turn around the forehand which itself turns around the exterior front feet, the left, if the haunches are displaced from left to right. It is important that the forehand move as stated, for should it be otherwise it could only be executed by a retrograde movement which would weight the hindquarters and impede their action.

It is thus seen that the movement of the forehand is reduced to a minimum; one front foot is even immobile and all the movement is executed by the hindquarters. Consequently the aids to be used must cooperate in placing onto the immobile front foot as much of the weight as possible, thus facilitating the movement of the other front foot and of the hindquarters.

Therefore, if it is desired to turn the haunches from left to right, the weight of the mass must first be placed on the shoulders by closing the legs and relaxing the fingers until the movement forward becomes imminent. The center of gravity is then prevented from further forward displacement and the weight is carried on the left shoulder by closing the fingers on the right indirect rein of opposition in front of the withers and on the left direct rein, care being exercised to hold the head and neck in the direction of the axis of the horse. While this is taking place, the left leg is moved a little in rear of the girth to displace the haunches to the right. The right leg is closed on the girth to maintain the forward displacement of the center of gravity, to prevent the horse from backing, and to stop the displacement of the haunches to the right when desired.

The seat is carried to the right to facilitate the displacement of the haunches.

Summing up, the aids to be used in executing the turn on the forehand to the left are the following:

a. Equal action of the legs to bring the weight onto the shoulders.

b. Action of the left direct rein and of the right indirect rein of opposition in front of the withers, in such a manner as to weight the left shoulder while maintaining the neck straight.

c. Preponderant action of the left leg, slightly in rear of the girth.

d. Slight displacement of the seat towards the right.

If, in spite of all precautions taken, the horse shows a tendency to back, this must be stopped by a more energetic use of the legs, even inducing forward movement if such is necessary. If this tendency persists, or is accentuated, the trot or gallop should be taken up at once. This fault is often due to the failure of the rider to *first of all advance the horse's center of gravity,* an error which should not be again committed.

The first few times the rider executes this movement, it is unnecessary for him to try to maintain the forehand strictly in place. As he becomes more skilled in the application of the aids he will apply them with the exactness necessary to obtain the immobility of the forehand.

This exercise is the commencement of numerous minor difficulties, but is very useful in teaching the rider to *measure exactly the intensity of his aids.*

32. THE ABOUT ON THE HAUNCHES (PIROUETTE).

The about on the haunches consists of causing the shoulders to describe a circle, or an arc thereof, about the haunches as a pivot. This movement is quite difficult as it requires that the hindquarters do no more than turn around one hind foot as a pivot, and leaves the displacement of the entire mass to the forehand. Consequently the rider should endeavor to sense what is taking place under him, in order to perceive the more or less marked movements of the hindquarters.

The rider proceeds towards the execution of this movement by requiring the horse to execute half turns, whose radius he progressively decreases. When this radius has become very short, the haunches often have a tendency to swing to the outside. The rider's outside leg should oppose this and cooperate with the inside leg to prevent the horse displacing the weight of his shoulders onto the haunches and backing.

Through the repetition of this exercise, the rider so decreases the size of the half turn that the haunches scarcely leave the track on which he is moving. The about is not obtained by increasing the tension on the reins so as to prevent horse from moving forward, and at the same time by accentuating the lateral movement of the forehand. The hindquarters must be immobilized and to accomplish this the outside leg prevents displacement of the haunches in that direction if the horse attempts such action, as he frequently does. The inside leg is held ready to intervene if the horse attempts to back. The hindquarters being thus immobilized and the hands continuing to assure the displacement of the shoulders, the latter turn around the haunches. This is the about on the haunches or the "Pirouette."

The rider, being at the halt and desirous of executing the about on the haunches from right to left, establishes contact with the mouth and bit in such a way that the weight is carried back towards the hindquarters without, however, causing the horse to back. This result, considerably more difficult than in the about on the forehand, is obtained by the effect of the hands in containing the impulsion demanded by a simultaneous and judicious action of the legs. A position sufficient for the time being is obtained when the horse's neck is elevated as much as possible without causing him to back or to get behind the bit. This result having been obtained, the left direct rein of opposition, and the right indirect rein of opposition in front of the withers, act so as to give the neck a slight bend to the left and to cause the circular displacement of the shoulders from right to left. The reins ought not to restrain the forward movement too much, thus provoking the horse to back, or too little, thus permitting the haunches to move forward in order to follow the forehand. In addition, the right leg intervenes when necessary to prevent the displacement of the haunches to the right.

Under these conditions the hindquarters, neither backing, advancing nor swinging to one side, remain in place while the shoulders turn around them.

Thus, the aids to be employed are as follows:

> *a.* Simultaneous action of the two legs, inducing the horse to move forward, and preparing him for the action of the reins.
>
> *b.* Left direct rein and right indirect rein of opposition in front of the withers, displacing the shoulders to the left and acting with such intensity that the horse neither advances nor backs.
>
> *c.* When necessary, increased action of the right leg to prevent the hindquarters swinging to the right.

If the rider has difficulty in executing this movement correctly, he may allow the hindquarters to advance slightly, but must not, in any case, allow the horse *to back*. The about on the haunches demands an accurate use of the aids and should only be attempted when a certain mastery has been obtained therein.

33. TO CADENCE THE TROT.

The trot is said to be cadenced when it is characterized by complete regularity of stride, the powerful extension of the hocks, and by such an independence of the diagonals that the horse seems to be received on one from the other in a complete state of equilibrium; and to change his support, not in order to maintain this equilibrium, but to advance. The gait becomes beautiful and gives the observer an accurate idea of the power of the horse. The maximum cadencing is obtained in the "Passage" where each pause on the supporting diagonal is well defined and marked. The "Passage," however, pertains to the domain of higher equitation. Cadencing the trot requires less skill and a less complete accord of the aids, and is an exercise that any rider with good hands can attempt.

To cadence the trot one diagonal must be caused to support the mass in a state of complete equilibrium in order that the other may be raised and extend its stride in complete liberty.

Let us take the case where the horse, at the trot, is about to receive his mass on the right diagonal. To obtain equilibrium on this diagonal, the fingers must be closed on the right direct rein of opposition, and the left indirect rein of opposition in front of the withers, and the right leg must act with greater force than the left leg, though both legs must be used strongly at each stride. The right diagonal thus supports the horse's center of gravity.

A slight yielding of the hand at this instant permits the left diagonal to advance itself. When this latter diagonal is about to receive the horse's mass, aids, just the reverse of the above, are used, and the mass is now taken up by the left diagonal as it previously was by the right.

The horse is thus sent from one diagonal to the other, and held in equilibrium on the supporting diagonal while the other makes its stride.

When the horse is sluggish or lacks suppleness, his trot can be cadenced by balancing him by executing the counter change of hands on two tracks. He is then forced to change direction, necessitating a shift in his equilibrium under the diagonal action of the aids, and becomes ready for cadencing his trot.

This requires a certain amount of refinement in the use of the aids, but the difficulty is materially lessened as the reactions are sensed.

When used by riders whose experience is not great it often results in a horse switching his haunches while executing this trot, due primarily to the loss of the effect of the outside leg. Excellent results may be obtained by using both legs equally and determining the cadence by the rein effects. This method is recommended for the majority of riders.

This exercise is a very important one in ordinary equitation, for a well cadenced trot is more easily extended. This is why it is dealt with here, and is the reason all riders are urged to cadence the horse's gaits, even if only to give them brilliance and extension. A few lessons suffice to give excellent results.

34. TO EXTEND THE TROT.

The extended trot, in which the strides are very long, is not necessarily a fast trot. To become such the strides must be repeated rapidly.

When the horse is accustomed to the cadenced trot, he can be very easily taught to extend it. The riders legs act as in obtaining the cadenced trot, but they must demand greater impulsion so that the extension of the horse's hind legs will push the mass a greater distance forward during the period of suspension. Collection should be greatly diminished or even

abandoned, the neck allowed the greatest extension possible, and the horse's mouth given a light support in order to give him confidence and to obtain a greater drive from his hind legs.

35. TO BACK.

a. The horse being in hand, a backward movement, from the halt, is started by closing the fingers of a fixed hand on the adjusted reins, accompanied by a slight shifting of the weight to the rear. This should not be a steady or continuous action, but is accomplished with quickly repeated applications until the horse takes one or two steps to the rear. After the horse commences his first step to the rear, the fingers relax and again close as the next step is demanded. If the horse backs too fast, or lowers the croup too much, the legs should act to counteract this tendency.

To halt, the action of the direct reins should cease. The weight should be carried forward and the legs used sufficiently to stop the backward movement. In training, the back should always be followed by moving two or three steps forward.

b. This movement, like the turn on the forehand, may cause the horse to *"get behind the bit"* and, besides, may produce serious injuries to the tendons and hocks of a young or under developed horse. Therefore, with this type horse, its use should be limited to that required to obtain the ready obedience of the horse in this movement. Backing is an excellent exercise in obtaining vertical suppleness in the horse, and the easy engagement of the hindquarters.

36. THE SHOULDER-IN.

Nearly all authorities agree that the Shoulder-In is one of the most useful exercises in obtaining suppleness and liberty of action of all parts of the horse's body. It is an exercise easy of execution and its results are excellent and quick. Many famous authorities recommend that when in trouble, revert to the Shoulder-In, relax the horse and start over again.

The correct execution of the right shoulder-in demands that the horse move freely forward on the line followed when the movement is demanded, or at an angle to the left of not greater than 30 degrees, at the gait and pace desired, with the entire body bent to the right, but with the most pronounced bend in the region of the poll and just in rear of the withers; the right feet carried across in front of the left feet; the jaw relaxed and accepting contact with the bit; the impulsion maintained or even increased, and reflected in the flexing of the relaxed jaw and poll.

A trained rider can tell by the feel of his horse whether or not he is placed properly and moving properly. For the inexperienced rider the following rule of thumb for the placing of the horse may be helpful:

(1) Right hind foot tracking the left fore. This is the most usual position for a shoulder-in on a straight line, but it may be equally correct with the right hind considerably to the left of the left fore.

(2) Full bulge of the right eye visible to the rider.

To obtain a right Shoulder-In:

a. Shorten the right rein.

b. Move the shoulders off the line previously followed by causing the horse to start to turn to the right, by means of the right leading rein assisted by the right leg in rear of the girth.

c. As soon as he is in this oblique and circular position, resume the original direction, maintaining the horse bent, by means of the right indirect rein of opposition acting in the direction of the left haunch, and the right leg in rear of the girth.

d. The left rein cedes, then limits the bend of the neck, and assists the right rein in carrying the shoulders forward and to the left, or blocks the left shoulder in case the horse attempts to escape the movement by sidling to the left.

e. The left leg is active on the girth, to assist in maintaining the impulsion and to check excessive displacement of the haunches to the left.

f. The rider's weight should incline slightly in the direction of movement; that is, toward the left shoulder.

For a good execution of the movement it is necessary that the hand which retains the inside shoulder act rather steadily on a fixed rein and not with traction from front to rear. Both legs must be energetic in their action.

The horse should be exercised frequently in this movement. The shoulders must be carefully alternated. Between each change he should be moved on a straight line for several steps. This suppling should be of short duration only. One should profit by the engagement of the hind legs to push the horse into a free extended trot after a period of work at the Shoulder-In.

This exercise gives the horse complete suppleness and greater freedom in all parts of the body. It brings about:

a. In the shoulders:

(1) Freedom, due to the rotation and elevation required of the right shoulder in carrying the right forefoot across in front of the left.

(2) Lightness and obedience to the hand, through the mutual reinforcement of the right rein and right leg.

b. In the haunches:

(1) Suppleness. The right haunch is lowered in passing the right hind foot in front of the left.

(2) Engagement, due to the carrying of the right hind foot forward under the weight of the body.

(3) Obedience to the leg, through the mutual reinforcing of right rein and right leg.

It puts the horse up to the bit because the leg acts in the same direction as the hand obliging him to accept the tension of the rein.

Suppleness of the spinal column, obtained by requiring the horse to move forward while in a bent and oblique position, which results in a constant play of the articulations of the spinal column.

Suppleness of the jaw and poll. The horse is less able to resist the hand by contraction of the jaw and poll when in the bent position.

The following are common errors in the execution of this exercise:

a. Neck bent too much, particularly just in front of the withers. Correct this by increased action of the left rein and the right leg. Avoid traction to the rear on the right rein. Right hand should act by fixity.

b. The horse over-flexes at the poll and his muzzle comes to the rear so that the line of the face is in rear of the vertical.

c. Jaw and poll stiff. This is usually caused by rough hands or by excessive demands on a horse not yet prepared for the movement.

d. Slowing down and turning to the right, while displacing the haunches to the left. This is caused by a loss of impulsion, due either to insufficient activity of the rider's legs, particularly the left leg, or to excessive resistance in the hands.

With a young, untrained horse, too much should not be demanded at first. He should be started in the movement first on a circle at the walk, and only a few steps at the shoulder-in required. When executed on the circle, the shoulder-in should have the effect of enlarging the circle travelled by the horse. If on the track to the right hand a favorable position for beginning a right shoulder-in is just after leaving the far corner and on entering the long side of the hall.

Only when the horse does the movement well at the walk on the circle, should he be advanced to the work at the trot and on the straight line. In other words, the work should proceed deliberately and progressively.

The shoulder-in is a difficult movement when accurately and precisely executed. However, it is possible to obtain the movement in less perfect form and for a very few steps at a time, near the beginning of the horse's training, due principally to the mutually reinforcing effect of the rein and leg on the side toward which the horse is bent. Therefore, it is a very useful training and suppling exercise for remounts, when used with discretion and with a knowledge of the objects sought and the mistakes to be avoided.

The shoulder-in is the only schooling movement where the position of the horse's muzzle does not determine the direction of movement.

37. WORK ON TWO TRACKS.

The rider, knowing how to effect the lateral displacement of the forehand and the haunches independently of one another, whether in motion or at the halt by means of ranging the haunches or forehand, the work of accurately displacing these two simultaneously should now be commenced, and is accomplished by the work on two tracks.

The horse is said to be moving on two-tracks to the right when he moves off his original line of march obliquely to the right, with the long axis of the horse remaining parallel thereto, and the shoulders and hindquarters following two parallel paths. The horse's head is turned slightly at the poll toward the direction of movement, and the shoulders slightly lead the hindquarters.

The outside feet pass over and in front of the inside feet. When the work on two tracks is executed properly, it is done with such harmony and symmetry between the different parts of the horse that his movements are made without loss of speed or of impulsion.

The aids employed in two tracking to the right are as follows:

By a light leading effect of the right rein, the rider turns the horse toward the right oblique. As soon as the horse has taken one or two steps in the new direction, the rider uses the left leg in rear of the girth to push the haunches to the right, and the left indirect rein of opposition in rear of the withers, to carry the horse's whole body to the right, while preventing an increase of speed. With the right leading rein, the rider continues to draw lightly the forehand to the right so as to set the head a little to the right and thus cause the horse to look in the direction of the ground he is to move over. The rider uses his right leg on the girth to maintain the forward movement, and inclines the weight of his body slightly forward and to the right to facilitate the movement of the horse in that direction.

By the combined use of the aids described above, the rider causes the horse to continue to move obliquely to the right front on two tracks. The tracks followed by the fore and hind feet make an angle of approximately thirty degrees with the original direction of march.

One of the most common errors committed in this exercise is that the rider, in two tracking to the right retards the movement of the shoulders by using the left direct rein, or the right indirect rein of opposition, in front of the withers, instead of assisting the movement by using the reins in the direction of movement. Although a movement resembling that of two tracks is readily obtained by such a method, it causes an entire loss of impulsion, as the aids used retard the action of the shoulders, which in any case move laterally with difficulty.

The direction in which the horse moves in two tracking *should not exceed an angle of 45 degrees to his original direction.* The horse's structure prevents the free movement of his outside legs if this angle is increased. If this is done, he will disrupt his equilibrium, become restive and may painfully injure his knees and cannons. Moreover, if the angle is exaggerated, the haunches will not be properly in rear of the shoulders, the forehand will not benefit from the drive of the haunches, and impulsion will be lost. Therefore, when the horse tends to exaggerate this obliquity, both legs should act energetically and with equal force to drive the horse in the direction indicated by the reins.

In commencing a new movement the rider should be satisfied with obtaining a few correctly executed steps, and should not prolong the movement. The study of the work on two tracks is productive of excellent results, as it trains the rider to feel his horse, and to coordinate his aids in accordance with the displacements which he cannot see but which, due to his equestrian tact, he can feel. His tact is developed thereby, and the aids are improved in accord and exactness.

The combination of the shoulder-in and the two track movement makes an excellent exercise to supple and cadence horses as well as to develop tact in the use of the aids for the rider.

This combined exercise is best obtained by going progressively from a right shoulder-in to a two track to the right, then after a few strides to a left shoulder-in and from this to a two track to the left. Each movement should be limited to only a few strides.

38. THE HAUNCHES-OUT.

Haunches-out is an exercise on two tracks with the horse's haunches remaining on the track. To execute it, the forehand must first be led to the inside, as in commencing a turn. At the instant the hindquarters are about to leave the track, the horse is moved along the wall by means of the aids described above for the movement on two tracks.

39. THE HAUNCHES-IN.

This exercise consists in moving the horse on two tracks with the shoulders on the track while the haunches move on an inside track.

To displace the hindquarters to the inside the same means must be employed as in executing an about on the forehand, with, however, a slowing up of the forehand. As soon as the horse is in the correct position to move on two tracks he is moved thereon, while continuing to use the outside leg preponderately, the inside direct rein and the outside indirect rein of opposition in rear of the withers.

As soon as the rider is able to two track sufficiently well on a straight line, he can perfect himself in this movement by holding the haunches in or out on the circle, half turn, broken lines, etc. This last movement is particularly useful, as it teaches the rider to reverse his aids rapidly and to give them at once the necessary force and intensity.

Both the exercises haunches-in and haunches-out are movements on two tracks in which the spinal column remains straight. They must not be confused with movements which move the haunches to the right or left and break the spinal column at the base of the neck. *This latter movement should be avoided and never used in training.*

40. GALLOP DEPARTS.

When the rider, as a result of a close study of the abouts on the forehand and haunches and of work on two tracks, has acquired the ability of maintaining his horse in equilibrium, he can be practically certain of obtaining the gallop depart with the lead desired. To take the gallop with the right lead, for example, the left hind leg must commence the movement by raising the entire mass by its drive; the right fore and right hind legs must also lead the left fore and left hind legs respectively, in order that the right lateral may constantly take its support in advance of the left lateral.

Therefore, if the left hind leg is weighted by the seat, it is evident that it is the member best fitted to raise the mass, since the right hind leg, being completely disengaged, has evidently but a slight effect thereon. If the reins weight the left shoulder at the same time, then the right fore leg, due to the shift in weight and to the drive of the left hind leg, is more easily able to extend its stride than the left fore leg. The right hind leg, put in motion when the mass is raised, necessarily leads the left hind leg which is still on the ground; the right lateral thus leads the left lateral.

It is evident, therefore, that if aids are used which weight the left lateral and cause the drive to come from the left hind leg, the gallop with right lead will necessarily be forced, especially if the right lateral is given a more advanced position than the left lateral.

This position can be obtained by a number of combinations of aids. The elementary method uses the outside lateral aids weighting the outside shoulder by turning the head and neck in that direction and by pushing the haunches to the inside. Another method employs the diagonal aids as explained below, and a third method employs the lateral aids on the inside. This latter method pertains to superior equitation and will not be discussed in this manual.

To obtain the gallop with the right lead, using diagonal aids, the following means should be used:

a. Weight of the seat on the left buttock thus charging the left haunch.

b. Left leg acting slightly in rear of the girth displacing the haunches to the right and consequently causing the right lateral to lead the left lateral.

c. Right leg acting on the girth to assist the left leg in producing the necessary degree of impulsion, and to prevent too great a displacement of the haunches to the right.

d. Right indirect rein of opposition in front of the withers, weighting and retarding the left shoulder, while freeing the right shoulder and allowing it to extend.

e. Left direct rein of opposition assisting the action of the right rein, and maintaining the head and neck straight.

The two reins must also oppose the passing of the impulsion to a certain degree, so that the excess impulsion may be used in lifting the forehand.

Then by a slight yielding of the hands, as the right fore leg begins to extend, the left hind leg pushes the entire mass into the gallop.

To facilitate the accuracy of the departs, the rider may commence by demanding them either in the corners of the hall, or at the completion of a circle, but always in such a manner that the horse takes the gallop *at the moment he returns to the straight line.* Under these circumstances, as in a turn to the right for example, the right lateral is in advance of the left lateral, thus placing the horse in a favorable position for taking the gallop with the right lead.

However, in this turn, the right shoulder and right haunch are more heavily weighted than the left members, which is an *unfavorable* condition for the gallop right. Therefore, the horse must be made to take the gallop at the instant he enters the straight line, for, although he is still curved to the right, the weight of the forehand and hindquarters must be moved to the left lateral in order to move out on the straight line. These conditions assure the gallop right.

Further, to make more certain of obtaining a smooth and tranquil gallop depart, the movement should be demanded from the trot, as the horse then is assisted by the impulsion already acquired. The legs then act less energetically, and the gallop depart is smoother.

As soon as the rider can habitually obtain the gallop departs at the completion of a turn, he may commence to demand them on a straight line from the trot, slow trot, and walk.

41. GALLOP DEPARTS WITH THE HORSE STRAIGHT.

In order to obtain gallop departs with the horse straight, (viz:—the horse's axis remaining exactly in the direction he is moving,) the rider's inside leg must act at the proper moment and with the intensity desired. In a gallop depart with the right lead, for example, the right leg, acting too late, allows the haunches to be displaced to the right, and, in acting too soon, does not afford the left leg sufficient time to set the horse for the gallop right, and consequently to lead off with this foot.

The right leg should receive the horse at the instant he starts this displacement to the right and when this attempted displacement has caused the right lateral biped to lead the left.

The horse must execute the gallop depart straight if his axis is to be straight while galloping. This position is indispensable for two reasons: first, if he gallops traversed, his hindquarters do not drive him exactly in the direction of march, and consequently a considerable degree of their effect is either lost, or without useful effect; secondly, the haunches, being curved in, make it difficult for the horse to change direction, and he may even refuse to turn by a further displacement of his haunches. Lastly, the freedom of his movements is diminished, and contractions result which hinder handiness and submission.

The rider should attempt, for the above reasons, to obtain straight and correct gallop departs. It is needless to state that this is difficult, but it is necessary. It may be said of a rider whose horses are always straight, especially at the gallop, that he has a precision in the use of the aids which enable him to meet successfully all the delicate situations encountered in ordinary equitation.

In gallop departs, there always arises the question of tact in determining the intensity of the aids. This varies with the horse and the pace. It should be obtained but not exceeded by the rider.

The expression "to lift the forehand" may be used because it is most important, but one which should not give a false impression. In fact, it is a common sight to see riders attempting to raise the forehand by using the hands or by tension of the reins. This latter effect is needless, for the rider does not raise the forehand; he causes the horse to do so without any expenditure of effort. He pushes the horse up onto the closed fingers, and further gives him a degree of impulsion which is not allowed to escape to the front. This force causes the horse's center of gravity to move slightly to the rear and causes the "lifting of the forehand," especially the foreleg which is freed of weight by the indirect rein of opposition in front of the withers.

42. THE FALSE GALLOP.

The horse is said to gallop "false" when, in making a turn, he leads with the outside instead of the inside foreleg.

The horse is prepared for work at the false gallop by galloping on broken lines, and by accentuating progressively the false turns involved in this movement. Following this, the figure of eight and serpentine are taken up.

The work on the figure of eight is commenced on a large eight, inscribed in the length of the riding hall, and executed only once in a period of work (two consecutive changes of hand). When the movement is executed without excitement the horses may be worked at this exercise for a longer period of time. When they gallop calmly, and well extended, the figure is made smaller, little by little. The suppling should be sought for by frequent periods of work on the large eight, rather than by an exaggerated closing of the figure; especially is this advisable in the case of cold-blooded horses.

Similarly, the serpentine at first consists of only one turn, which is made gradually smaller as the horse progresses.

This work leads the horse himself to modify his balance, and it gives the alternation of extension and engagement which is sought in all gymnastics. It is also an excellent exercise to extend the horses' head and neck as the horse is obliged to use the muscles of the neck more than he normally would in order to extend the outside foreleg.

In executing the false gallop to the right, the right indirect rein of opposition in front of the withers and the right leg predominating keep the horse from changing to the inside lead.

43. TO CADENCE THE GALLOP.

The gallop is a gait susceptible of a speed that, even when it is not exaggerated, may render the horse difficult to control. It should therefore be cadenced.

To do this, advantage should be taken of the fact that the horse, at the second beat of the gallop, extends his neck accompanying and assisting the movement of the foreleg about to execute the third gallop beat. If the fingers are closed so as to oppose this extension of the neck, the horse becomes collected, and his center of gravity moves to the rear. The energy of this extension depends on the impulsion; therefore, the more it is desired to slow the speed of the gallop, the greater should be the impulsion, and consequently the more active should the legs be to maintain it. By this means the speed of the gallop can be slowed until eventually the gallop in place is obtained, without the gait being lost.

Many horses naturally gallop slowly, but stretched out and without impulsion. On these the action of the legs should be severe to awaken them and to enliven their gait, and usually it is best to gallop them at speed out-of-doors.

When the horse takes a cadenced gallop, his neck is high, his hindquarters are engaged, his jaw and poll are flexed and relaxed, and he is easy to control.

Collection is more easily obtained and maintained on straight lines, and on wide turns, than on changes of direction and short turns. For this reason the rider must first of all seek to obtain collection while galloping on a straight line, and then attempt to maintain it on circles whose radii are not reduced until the rider can remain on them without provoking contractions or loss of impulsion.

Finally the rider attempts a more difficult task, which increases the accord and finesse of the aids, viz: two tracking at the gallop with the horse collected. The same aids are used as at other gaits. In the two track to the right, for example, the left leg directs the haunches toward the right, the left indirect rein of opposition in rear of the withers acts similarly on the shoulders, the right direct rein bends the poll slightly toward the right, and the right leg regulates the displacement of the haunches and keeps the horse up on the hand. The rider's weight is carried slightly to the right, as much to facilitate the displacement of the haunches as to conserve a steadiness which would otherwise become difficult.

44. THE CHANGE OF LEADS.

Due to the mechanism of the gallop the horse can turn easily only on the inside foot. A smooth turn with the horse galloping false requires an amount of training which does not pertain to elementary equitation. Therefore, a rider, galloping with the right lead, for example, must come to the trot, or change leads, if he desires to turn sharply to the left. If the lack of time, or the desire for a less elementary equitation, prevent coming to the trot, a change of leads must be obtained. This movement consists of bringing about, in the same stride, a complete and instantaneous inversion of the equilibrium and mechanism of the horse's legs.

To bring about this change, there must be an analogous inversion in the aids. In other words, if a change of lead from gallop right to gallop left is desired, the following changes must be made:

> (a) The right leg must take over the preponderant action of the left leg.
> (b) The left leg must act only to push the horse into the bit, causing him to receive the indications of the reins and to hold him straight.
> (c) The left rein must act as an indirect rein of opposition in front of the withers.
> (d) The right rein must act as a direct rein.
> (e) The seat must be shifted from the left to the right.

This five-fold operation must be executed with harmony, tact, and decision, in order to obtain a precise and regular movement.

The rider should attempt this progressively and not try to accomplish it all in the first few trials. For example, being on the track on the right hand at the gallop, the rider executes a half turn coming to the trot upon completion of the half circle, in order to return to the track. Upon reaching the track, the rider takes up the gallop with the left lead. Between the moment when the trot was taken and the gallop resumed, five or six yards were covered at the trot, during which space the rider has time tranquilly to reverse his aids. This exercise is repeated first on one hand, then on the other, until the rider feels master of his aids at the moment the gallop is resumed. Then he increases the difficulty of reversing the aids by coming to the trot a little later, thus decreasing the amount of time available for such action. Gradually this can be done in two strides, then in one. Finally, when he is able to change the horse's equilibrium smoothly in this short interval, he requires the horse to change from the right to the left lead without the intervening time at the trot. This is the change of leads. The aids used in effecting this, have already been described. When the change of leads is executed correctly at the end of the half turn, the rider then endeavors to obtain it on a straight line.

As in the gallop departs, it is unnecessary, in commencing the study of the change of leads, to require the horse to be completely straight during its execution. However, as accuracy in the use of the proper aids is required, the haunches should be held exactly in rear of the shoulders. The reasons are similar to those demanding gallop departs in a straight line, and, in order to change leads with the horse straight, the interior leg should act as stated for gallop departs. Finally, the horse should be kept collected, which is done by the timely and exact application of the aids, in the case of a trained horse. Should the horse be only partially trained, the effect of these aids will undoubtedly improve him and prevent contractions.

The change of leads thus obtained is the most difficult movement in ordinary equitation. It completes the list of exercises which train the rider in those qualities necessary for riding and controlling his horse in the various circumstances encountered in outdoor riding.

45. OBTAINING BALANCE.

In the extensions, collections and changes of gaits which have been studied, there was always an underlying obligation on the rider to avoid contradiction of the aids by inclosing the horse between active legs and hand. The rider must observe that only by an exact balance of hand and leg action is the horse's impulsion carefully preserved and his submission retained. But as training progresses towards perfection, these actions tend to approach each other until finally they seem to merge into one another. The horse, thus balanced between hand and legs, moves with high strides in a sort of equilibrium. This is nothing else than collection while marching, which has for its object the shortening of the animal's base of support so that he then works on a *short base*. This accentuates his mobility while necessarily reducing his speed.

When it is desired to return to an extended gait it is necessary only to abandon the collection and to allow the impulsion to fall; then the neck stretches out, the gait is extended, and the horse works again on a *long base.*

The object of the changes of gait and the increase and decrease of speed, is to oblige the horse to work on a short or long base according to circumstances, and to accustom him to pass from the most collected work to the most extended, as in outdoor equitation. In a word, it is to put the horse in balance.

46. REMARKS ON THE GAITS.

From the mechanism of the walk, trot, and gallop, one may make some observations which are very useful in training.

It is very necessary that the rider should know how to take a certain gait, to maintain or to change that gait. In the walk and trot, the lateral bipeds move in parallel planes; in the gallop, on the contrary, the horse tends to travel with one haunch slightly to the side. This suffices to show the rider that in the walk and trot the horse should be held *absolutely straight,* while in the gallop he should yield towards one haunch *very slightly.*

With a young horse, therefore, whenever the rider wishes to take the gallop he should give him this natural position. On the other hand, when he wishes to pass from the gallop to the trot, or walk, he should straighten his horse.

The rider should also be able to obtain, when desired, those gaits which allow the horse to cover the most ground with the least fatigue. For the horse to expend a minimum of energy, the impulsion must all act in the direction of the movement.

The horse travels low when he moves with his neck extended so that its direction approaches the horizontal. The muscles of the neck in this position draw the forelegs forward and not upward. The hindquarters under the action of the rider's legs may easily engage under the mass because the position of the neck allows the spinal column to arch and then extend in the direction of the movement. This position will therefore favor speed, and all the horse's efforts will tend to produce forward movement with a minimum of fatigue and expenditure of energy. It is this position which the rider should seek to impose on the horse whenever he wishes to work on a long base.

A free walk, extended without exaggeration, constitutes a rest. It is therefore an excellent reward, and at the disposal of the rider to show the horse his satisfaction with a well-executed movement; it should therefore be used frequently.

Furthermore, at this gait the seat is steady and the rider is in full control of his aids, therefore, he should use it when correcting faulty positions of the horse, and when giving him the proper position preceding a new movement. The horse is more apt to obey in proportion to the clearness of the effects of the aids, and he is better balanced; therefore, at the walk he is in the best possible position to receive his lessons, and each new movement or position should be taught at the walk before proceeding to the more rapid gaits. However, this gait is slow, and there is little gymnastic effect on the horse's joints and muscles, especially in lateral movements.

At the gallop, an excellent exercise for the spinal column may be practiced by extending and collecting the gait on straight lines.

The movement on two tracks at the trot is a suppling exercise which becomes more effective as the impulsion increases and the gait is extended. But in order that the left members, for example, shall open widely toward the left and the right members shall pass in front of them, the horse must move with long strides and the speed must be maintained, otherwise there would be no deep work of the muscles and no unusual play of the joints; in other words, neither suppling nor impulsion. Work on two tracks at the gallop has, however, no particular useful effect, since in this movement the horse moves by a series of bounds parallel to himself and without crossing his legs, so that he expends no great effort.

47. EXTRACTS OF THE GENERAL REGULATIONS OF THE FEDERATION EQUESTRE INTERNATIONALE: SCHOOLING DEFINITIONS.

a. The following are definitions of gaits and schooling movements published in the General Regulations of the Federation Equestre Internationale, to assist riders in the proper presentation of their horses in the schooling tests prescribed by that organization.

Although the passage, piaffer, demi-pirouettes, and pirouettes at the gallop, are movements of higher equitation, they are reprinted here as a matter of information.

(1) At the halt, and in all his work, the horse should be "In hand."

The horse is said to be "In hand" when his hocks being in their place, the neck more or less placed according to the rapidity of the gait, the head fixed and the mouth relaxed, he does not offer any resistance whatsoever to his rider.

(2) At the *Halt* the horse, standing squarely on his four legs and immobile, should be ready to move forward at the slightest call of the legs.

(3) The *Free Walk* is a frank, extended and relaxed walk. The rider allowing a great amount of freedom to the head and neck, (without ever losing contact); the horse travels lightly, but calmly, with equal and deliberate steps.

(4) *At the Collected Walk,* the neck is raised and arched, the head assumes a position approaching the vertical, the hind legs are engaged. The gait becomes shorter, but the mobility of the horse is greater.

(5) *The Ordinary Trot* (natural or medium) is an intermediate gait between the extended trot and the collected trot. The horse moves forward freely without crossing, stretches the reins lightly, in a balanced and relaxed attitude. The strides should be as nearly equal as possible; the *rear* feet follow the front feet accurately.

(6) *At the Extended Trot,* the horse extends his action; the neck stretches; the shoulders, driven energetically by the haunches, gain ground to the front without sensibly increasing their height.

(7) *At the Collected Trot,* the neck is elevated, permitting the shoulders to act with greater liberty in all directions; the hocks engaging under the mass, keep up, despite a reduced speed, the energy of the impulsion. The horse takes shorter steps; but he is more mobile and lighter.

At the ordinary trot and extended trot, the rider generally rises to the trot (posts).

At the collected trot he sits the trot.

(8) *The Ordinary Gallop* is an intermediate gait between the extended gallop and the collected gallop. The horse, perfectly straight from the head to the tail, moves freely in his natural balance.

(9) *At the Extended Gallop,* the neck extends; the muzzle is carried more or less forward; the horse increases the length of his strides without losing any of his calmness or lightness.

(10) *At the Collected Gallop,* the shoulders, well-disengaged, are free and mobile, the haunches are active and regular in their action; the mutability of the horse increases, without loss of impulsion.

(11) At all gaits, a light mobility of the jaw, without nervousness is a guarantee of the obedience of the horse and of the harmonious distribution of his forces.

(12) The changes of gait and pace should always be distinct, rapid and yet smooth; the previous cadence is maintained to the point where the horse takes the new gait—or halts.

(13) *In the transition* from backing to a movement forward, the horse should not show any sign of halting.

(14) *In the changes of direction,* the horse should, according to the School followed, either remain straight or bend lightly on the arc of the circle which he describes.

(15) *In the work on two tracks,* the head, neck and shoulders should always lead (precede) the haunches; the horse ought not to show any trace of slowing; a very slight bend, to permit his looking in the direction in which he travels, adds to his grace and aids in freeing the outside shoulder.

(16) *In the repeated counter-changes of hand* (zigzag), the attention of the judges is directed on the attitude of the horse, the crossing of his members, the precision, the suppleness and regularity of his movements. At the gallop, moreover, the number of strides executed is taken into account.

Any abrupt movement at the moment of changing direction is a fault.

(17) *In the half-turns on the haunches at the walk and demi-pirouettes at the gallop* the shoulders describe a half-circle about the haunches: they begin their movement without any halt, at the moment when the hind legs cease to advance, and move forward, without any halt, as soon as the movement terminates.

(18) At the gallop, the horse changes leads " in the air" in a single stride and while advancing. The change of lead is said to be "in the air" when it is executed during the short time of suspension which follows each stride of the gallop. The horse should remain straight, calm and light.

(19) To execute the *Serpentine,* the rider commences his first loop by progressively gaining distance (moving away) from the short side; he terminates his last loop by progressively nearing the short side in the opposite end.

(20) The "Passage" is a collected and shortened trot, very elevated and very cadenced. It is characterized by the more accentuated flexion of the knees and the hocks, and by the gracious elasticity of the movements. Each diagonal, well united, is alternately elevated and put down, in the same cadence, gaining little ground to the front and prolonging the suspension.

The toe of the foot of the fore leg in suspension should be elevated about to the height of the *middle of the cannon* of the fore leg in support—the toe of the foot of the hind leg in suspension should be elevated only a *little above the fetlock joint* of the hind leg in support.

The same Passage cannot be demanded of every horse. According to their conformation and their temperament, and also according to the energy of their impulsion, some horses have a more rounded and larger action, others quicker and shorter—but rocking the haunches is considered a fault.

(21) The "Piaffer" is the "Passage" in place, the members in suspension elevating a little higher than in the "passage."

(22) The Pirouette at the gallop is an about on the haunches, with a radius equal to the length of the horse. The shoulders describe a circle around the haunches, which remain at the center of the circle and serve as pivot. The hind legs continue the gallop in place, elevating and coming down almost in the same place, at the same time pivoting in order to follow the movement of the shoulders.

The horse should turn without abruptness, maintaining the same cadence and the same impulsion.

He should not take a continued support on one of his hind legs; in this case, he would cease to be at the gallop.

(23) All movements should be obtained without any apparent action on the part of the rider. The rider should be seated upright, the loins and hips supple, the thighs and legs fixed, the upper part of the body easy, free and straight.

The use of the voice in any manner whatsoever, clucking with the tongue, once or several times, are absolutely forbidden. They cause obligatorily the mark zero for any movement where they have been used.

Riding with the reins in both hands is obligatory, except in the movements where it is specified that the reins should be held in one hand.

48. MASTERY OF THE AIDS.

However theoretically precise the effects of the legs and reins may be, they can have practical utility only if the aids which produce them are perfectly disciplined and submissive to the will of the rider. It is not sufficient to know these aids, it is also necessary to be master of them.

If the horse does not submit to the requirements of the rider, in the majority of cases it is not due to the ignorance or bad will of the horse; but it is because the weak or incoherent application of the aids do not indicate the desired movement.

Coordination and independence of the aids are obtained by *controlling the reflexes.*

If young riders are ordered to act with the left leg alone, the right leg nearly always flies out an equal amount. This one example gives an idea of the great amount of work necessary to control the muscles so as to enable the rider to employ them for a definite useful purpose according to the rider's will.

Without dwelling upon the causes of what is commonly called "awkwardness," it is seen that the role of the instructor includes the bringing about and multiplying the occasions which the young rider has for using the proper aids correctly, first employing them singly, then in combinations.

The rider holding the reins separated in the two hands is directed to utilize in simple movements, such as the passage of corners, moving by the flank, and circles; first, the opening rein effects, then, the bearing rein effects and, finally, the effects of opposition, abandoning completely the rein which does not determine the movement.

Example:

"By the right opening rein, by the right flank;"

"By the right bearing rein, circle to the left;"

"By the left direct rein of opposition, half turn to the left."

He is then taught, by composite movements, to substitute the effect of the opening rein for the effects of opposition, or the indirect effects for the opening effects, etc.

Example:

The platoon marching to the left hand, the instructor will command:

"Half turn in reverse, leave the track by the bearing rein." "Right indirect rein of opposition, in rear of the withers, on two tracks on the diagonal."

"By the right opening rein, right about." Or again, the platoon marching to the right hand: "Half turn, by the right opening rein, right about." "By the left indirect rein of opposition, in rear of the withers, on two tracks on the diagonal," then, immediately, "By the right bearing rein, by the left flank, etc." The rider will have been shown in the first example that the right rein has been able to produce three effects *according to the different* directions given it; in the second, he will have learned to substitute rapidly the action of the left hand for that of the right and back to the former.

When this practice of one hand alone is well understood and executed, it is necessary to learn, by the same movements, and with the same progression, to act with both reins, but having them in accord, the hands acting, resisting or following according to circumstances.

In the end the movements must be rapid and complicated, such as the broken line, serpentine, to give the rider decision and agility.

Concurrent with these increasing difficulties the instructor should make sure that the riders use properly the prescribed actions of hands and legs and that they realize the effects produced. He indicates to them the positions desired for the head and neck, the dangers to avoid, and by constant criticism corrects every fault committed.

The student will thus come to discern the muscles which should act in the execution of the different prescribed movements, to isolate them, to put them into action, and to increase the power and rapidity of their actions. By experience, then, he will have only to acquire the habit of *true* and *timely* action to be in full possession of his powers as a rider and to be able to overcome all difficulties.

The rider should be alert to act with only the parts necessary to accomplish the result, for correct execution is prevented by the involuntary action of certain of the rider's parts which may occur without his knowledge. For example, the body should not be displaced when he uses the legs or hands; again, the knees should not be displaced when he uses the legs.

It is very essential also not to draw the right leg close if he wishes only to use the left, and similarly not to make use of the left when only the right is required, for the horse will not know what is asked of him. It is necessary to teach each man who mounts a horse the effects that are produced by each leg singly and when used in accord together. It is no less important to know the effect produced by each rein of the snaffle and curb, for often the rider employs the left when he should employ the right, and the right when he should employ the left, and often both when he should employ only one.

49. EQUESTRIAN TACT.

When the rider knows the means of control and is master of them, he has only to apply them with tact.

It is solely the application of the means of control which decides and regulates the movement, and directs it towards the accomplishment of a desired purpose.

Practice in the use of the aids gives birth to the *feel of the horse* and *equestrian tact*.

The feel of the horse enables the rider to judge the degree of submission or of resistance of his mount.

Equestrian tact regulates the degree of force used by the rider. It leads him to determine the effect to produce, the intensity of that effect, and the exact moment to produce it. It enables him to conquer resistances, or at least to forestall them.

The agents of equestrian tact are the legs and the hands.

a. Tact of the legs.

The legs can act only in one direction. In their use, then, there is only a question of intensity, which the aid of the spur renders more powerful. From the study of the mechanism of the gaits, the rider, by his seat, can have a certain feeling of the movements which constitute the raising, suspension, and planting of the feet; he can profit by this to hasten or retard their play, interrupt their combinations, and hence to correct or modify the gaits.

b. Tact of the hand.

The study of the action of the reins has determined their theoretical effects, but these effects may produce very different results according to the qualities of the hand which provokes them.

Finesse in the use of the hands is the most difficult part of horsemanship to master.

The qualities of a good hand are steadiness, lightness, softness, firmness.

To have a *steady hand* does not mean that the hand shall remain immovable; it should, on the contrary, move up, down, to the right, and left, according to need, but in the execution of this, it should be free from all involuntary or useless movement.

Steadiness of hand is the first quality to be sought, and is the most important of all, for without it the others cannot be fully developed. The unsteady hand can have neither lightness, softness, nor firmness; its indications are uncertain and the most attentive horse cannot obey its incoherent action.

The *light hand* maintains the merest contact with the horse's mouth.

The *soft hand* gives support.

The *firm hand* gives a frank, decided bearing.

The hand should know how to resist authoritatively when necessary, but should yield as soon as the resistance of the mouth disappears, and should return to softness which is always the bond or union between lightness and firmness. It is in this sense that a good hand has been defined as a "force in the fingers equal to the resistance of the horse, but never greater."

Actions of the hand vary in extent and intensity with the degree of training of the horse. At the beginning, the forearm, wrist, and hand participate in the action of the rein aids. *With a trained horse, however, it is only by a closing more or less energetic, or by an opening, more or less complete, of the fingers, that the rider transmits his will.* The pronounced action of the hands necessary with the green or partially-trained horse are through education, finally succeeded by effects of mere indication.

Horsemen are prone to forget that the technique of hands is always subject to improvement and at no time are they ever completely made. Progress in their skill is without limit.

To sum up, equestrian tact consists in choosing the correct determining and regulating aids, in assigning to each of its proper action, resistance, or passivity, and then by means of the aids, causing the effect to fall upon the *point selected,* (keeping in mind the seats of resistance which are the poll and jaw, shoulders and haunches) and as nearly as possible at the *instant desired,* so as to take advantage of the laws of balance and locomotion.

The role of the instructor is here much restricted because, not riding the horse himself, many resistances escape his observation. The pupil must, therefore, redouble his efforts to be honest with himself as to his faults. If he does not judge his own actions properly, he will make no progress. It is practice, founded on sound principles that should be his real teacher.

CHAPTER VIII
RIDING HALL MOVEMENTS AND CUSTOMS

50. Riding Hall Movements 113
    a. To march to the right (or left) hand 114
    b. To take track without regard to distance 114
    c. To take the track with fixed distances 114
    d. To ride at will 114
    e. To ride in 115
    f. By the right (or left) flank 115
    g. Right (or left) oblique 115
    h. The about 115
    i. To pass from the front to the rear of the column .. 116
    j. Work by threes 116
    k. The half turn 116
    l. The half turn in reverse 117
    m. To change hands 118
    n. Circling individually 118
    o. The figure eight 119
    p. Broken lines 119
    q. Serpentine 120
51. Customs and Etiquette in the Riding Hall 121

50. RIDING HALL MOVEMENTS.

The following subparagraphs give a few of the rules and exercises employed in class work to teach the rider the application of the aids to obtain certain desired effects of action from the horse:

a. To march to the right (or left) hand.

The rider is said to march to the *right hand*, if, being on the track, the center of the riding hall is on his right; to the *left hand*, if the center of the riding hall is on his left.

b. To take track without regard to distance.

There should be much work on the track by a class without regard to distance. Individuals are then absolutely independent of each other, the only obligation being to maintain the gait and to march to the hand directed when on the track. If the column becomes crowded, individual riders are permitted to turn out at will, move straight across the hall and take the track at a convenient interval on the opposite side or at the rear of the column.

When riding on the track without regard to distance, commands are not necessarily complied with at once; each rider, however, conforms to them as quickly as possible, considering his place on the track and the position and readiness of his horse for properly executing the movement. Each rider should choose his ground so as to avoid interfering with his neighbor. Under such conditions, thoughtful deliberation in using the aids and exactness of execution of the movement by individuals are the points to be demanded by the instructor, and not merely that all the riders shall begin or end the movement at the same moment.

c. To take the track with fixed distances

A designated rider takes the track, if not already there, at the gait ordered. The other riders move by the shortest line at the ordered gait, or increased gait if necessary, and place themselves one behind another with the indicated distance from head to croup. All then ride at the gait and at the distance ordered.

In the riding hall the distance from head to croup, when riding with fixed distances, is habitually four feet. Out of doors, however, greater distances should frequently be used such as 40 yards between riders. Such out of door exercises are very useful for training men to control the speed of their horses and to maintain a fixed distance from a rider or group of men preceding.

d. To ride at will.

The riders scatter out in the interior of the riding hall or over a designated area if out of doors, and riding at the gait ordered, execute on their own initiative such movements and exercises as the instructor may have specified.

When riders meet, each keeps to the right.

Horses and men alike seek companionship; they prefer to move in a body and do not like to act alone. Although they require collective training, they need far more to be taught to go ahead with confidence and courage individually. For that reason, riding *without regard to distance* and *at will* should be the rule both indoors and outdoors during instruction in equitation and riding *with fixed distances should be the exception.*

e. To ride in.

The riders moving at the gait at which they were riding or at the designated gait and by the shortest line, group themselves about the instructor. Such oral instruction, explanation or orders as may be appropriate are then given.

To resume the ride, the instructor gives the command or direction for the exercise desired. The riders move at the designated gait and by the shortest line to their places.

f. By the right (or left) flank.

Each rider moves his horse forward, and at the same time, uses the leading rein (or the bearing rein) in such a manner as to turn the horse to the right. The turn is made on an arc of 90°. The radius of the arc, when the horse is at the walk, should be 2 yards; when at the trot, 4 yards; when at the gallop, 6 yards. During the movement, the body of the horse should follow the arc on which he is turning. If, however, the horse prefers to carry his haunches somewhat to the inside of the curve, especially when at the faster gaits, the rider should not oppose it. But if, on the contrary, the horse carries his haunches to the outside of the curve, his balance is adversely affected, and the rider should oppose such action by using the left leg more strongly and slightly in rear of the girth.

In the first lesson in turning by the flank the rider uses the leading rein. Later he is taught to use the leading rein assisted by the bearing rein; later the bearing rein alone, first with the reins in both hands, then in one hand.

g. Right (or left) oblique.

This movement is executed in the same manner as *By the right or left flank*, except that each rider turns his horse through an arc of only 45°.

h. The about.

Each rider turns his horse until he finds himself facing in the opposite direction and then moves to his new front.

The turn should be made on the arc of a circle whose radius is 2 yards at the walk, 4 yards at the trot, and 6 yards at the gallop.

i. To pass from the front to the rear of the column.

The leading rider leaves the column by the right or left about, according to whether he is marching to the right or left hand, moves parallel to the column and enters it again at the rear by another about.

The riders in succession execute the same movement at the command *Next*, which is repeated by the instructor until all the riders have passed from front to rear.

The instructor may vary this exercise; for example, he may cause the first rider to move straight forward at the increased gait and, passing entirely around the circuit of the track, rejoin the column at the rear.

j. Work by threes.

The riders being on the track, on the right or left hand, with four feet distance from head to croup, as the head of the column passes across an end of the riding hall, the first three riders at the head of the column execute *by the right flank* and proceed straight down the length of the hall. They are followed by each successive three in the column who turn to the flank on the same ground. In each three the center rider is the guide. Upon arrival at the opposite end of the hall the leading three take the track to the hand to which they are ordered.

When the riders are disposed by threes in the riding hall, the instructor should cause them to halt, to move forward, to change gaits, to turn to either flank, and to execute the right or left about. Each rider, in addition to executing all movements correctly, should be required to preserve his place, checker-board fashion, with reference to his comrades. Thus the conditions of the exercise demand of the rider increased mental alertness and physical energy, combined with an effective use of the aids. Otherwise, he cannot keep his place.

k. The half turn.

The half turn consists of an about followed by an oblique. (See Fig. 11).

Being on the track to the right (or left) hand, at the walk, slow trot or trot, each rider describes a right about and then by the right oblique regains the track. The about should be made on the arc of a circle whose radius is 2 yards at the walk, 4 yards at the slow trot, and 6 yards at the trot, or to some line designated by the instructor, such as dot, white line and cross.

At the moment the horse starts the right about, the rider increasing the action of the left leg and left bearing rein, causes the horse to turn slightly on the haunches.

This movement being a turn slightly on the haunches, it should be much practiced, at the walk and slow trot, as it is a preliminary exercise to the about on the haunches.

Fig. 11.

l. The half turn in reverse.

The half turn in reverse consists of an oblique followed by an about. (See Fig. 12).

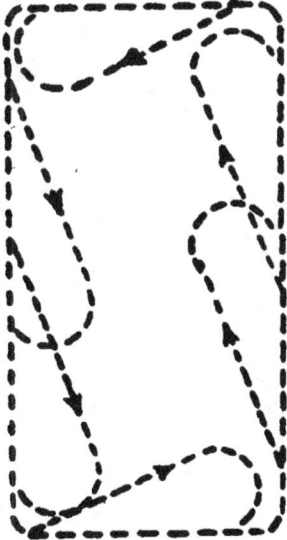

Fig. 12.

Being on the track to the right (or left) hand, at the walk or slow trot, each rider turns to the right oblique and then, at the proper time, by a left about regains the track.

At the moment the horse starts the left about the rider increasing the action of the left leg and left direct rein, causes the horse to turn slightly on the forehand.

This movement being a turn slightly on the forehand, it should be practiced only at the walk and slow trot. At the trot, the horse is often not well collected for executing the movement; moreover, the rider, rising to the trot, is not well situated for using the aids in making the turn.

m. To change hands.

The leader of the column, after having passed the far corner of the short end of the riding hall and marched a horse-length on the long side, directs himself toward the diagonally opposite corner so as to take the track to the opposite hand at about two horse-lengths from the corner. Those in rear follow in trace.

n. Circling individually.

The riders being on the track, at the walk, slow trot, trot, or canter, describe a complete circle tangent to the track and retake the track at the point where each left it. (See Fig. 13). The radius of the circle should be 2 yards at the walk; 4 yards at the slow trot and trot; 6 yards at the canter.

Fig. 13.

Each rider rides his horse with great exactness on the circle. He should endeavor to use the aids correctly in maintaining the gait, controlling the direction of his horse, and holding him with his spine bent to conform with the circle.

o. The figure of eight.

This exercise consists in describing, individually, the figure of eight, at the walk, slow trot, or trot.

In the beginning the figure should be of large dimensions, as for example, extending across the hall with the ends of the figure tangent to the long sides. Later, as the riders become more skillful in the use of the aids, the dimensions of the figure should be gradually reduced until finally it is approximately as shown in Fig. 14.

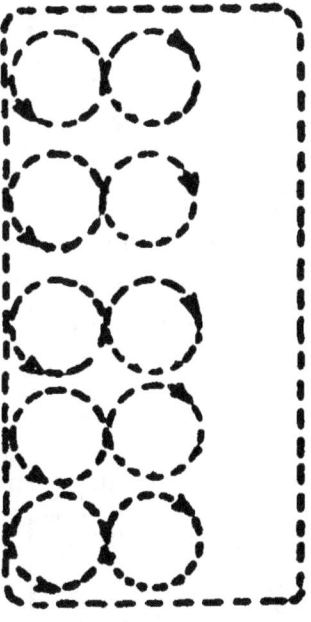

Fig. 14.

At the trot the posting diagonal is changed at the completion of each full circle so that the rider will post on the outside shoulders in describing each circle.

p. Broken lines.

The broken line consists of departures from and returns to the long side of the hall by obliques executed on the arc of a circle. As the leading rider enters the long side, he leaves the track by an oblique and, having marched the indicated distance to the center or to some designated line parallel to the long side of the hall, returns to the track by another oblique. This is repeated the number of times specified; at the other end of the long side he retakes the track. (Fig. 15a) Those in rear follow in trace.

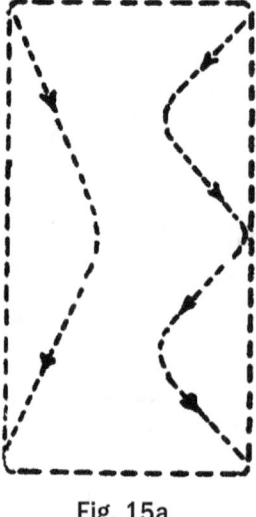

Fig. 15a.

This exercise in the riding hall teaches the rider to apply both his leg and rein aids *tactfully* in order that his horse may execute smooth changes of direction with lightness, suppleness of the spine and engagement of the haunches. This exercise is most useful as a preliminary step in maintaining a false gallop [counter-canter].

q. Serpentine.

The serpentine consists of successive abouts executed as indicated in Fig. 15b.

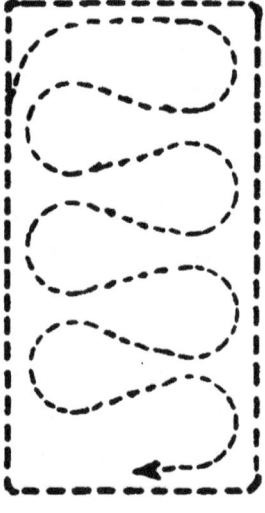

Fig. 15b.

Being on the track at the walk, slow trot, or trot, as the leader reaches the far corner of the short side designated by the instructor, he leaves the track and follows, by a continuous succession of abouts, a line as indicated in Fig. 15b. Upon reaching the opposite end of the riding hall, he retakes the track to the hand as indicated by the instructor. Those in rear follow in trace.

Riders should execute broken lines and serpentines as individuals; they should not follow too closely in the trace of the man in front. The riders being thus disposed, each turn is made by the horse in exact obedience to the aids as applied by the rider, and with the same object in view as in *p*.

51. CUSTOMS AND ETIQUETTE IN THE RIDING HALL.

It is apparent that when several riders are working horses in a small enclosure such as a riding hall, unless each shows consideration for the rights of his fellow riders, annoyance, confusion, and even serious accidents may occur. Accordingly, in all well-conducted riding halls, it becomes necessary to enforce certain rules of conduct. It may not be amiss to include here the following customs and rules of etiquette observed in military riding halls.

a. When the riding hall is being used according to a prepared and published schedule, be discreet about asking to enter if your presence might interfere with or annoy those occupying it.

b. Call "DOOR" loudly enough to be heard anywhere in the hall, and then wait for the reply "COME" before opening the door or entering. In the absence of an attendant at the door, be sure to close it on entering; if the hall is occupied, close the door upon departing. Enter promptly when the door is opened for you and move at once toward the center of the hall, looking sharply about to avoid getting in the path of someone who may not be able to avoid you. Enter either mounted or dismounted; if dismounted, move at once to that part of the hall least occupied, usually the center, and then mount.

c. After mounting and while warming up your horse at slow gaits, remain off the track and avoid those moving at faster gaits.

d. If, for any reason, you wish to halt, move toward the center of the hall.

e. Avoid blocking the track while talking to persons in the gallery or outside.

f. Keep to your right in passing all riders approaching from the opposite direction.

g. Keep to the side that is away from the wall in passing all riders going in the same direction you are.

h. Do not stop nor turn sharply without first looking behind you to see if the way is clear.

i. If someone is improperly blocking your way, call "TRACK" loudly enough to be distinctly heard.

j. If there is jumping in progress, it is inconsiderate of you to get in such a position as to cause the person jumping anxiety or fear. The one jumping should not be expected nor required to look out for others.

k. Horses led in the hall for exercise should be led on the side next to the wall to prevent them from whirling and kicking passers-by.

l. Unless occasion demands, longeing should not be permitted in a crowded riding hall. This applies as well to any form of dismounted exercise with resisting, mean, or vicious horses.

m. Manage your horse so as not to be a nuisance to others. It is irritating to have someone ride up on your horse's heels; the person who does so is also in danger of getting kicked. Do not ride your horse in such a manner so as to irritate or excite other horses.

n. If riding to the left hand at increased gaits, stay on the outside track. If riding to the right hand at increased gaits while others are moving in the opposite direction, stay on the inside track. If all are moving in the same direction, the outside track may be used while moving to the right hand.

o. Time your entrance into or departure from the riding hall so as to cause least inconvenience to others.

p. The management of any riding hall is governed to some extent by local conditions; furthermore, not all halls are managed in the same way; hence, make it a point to adhere strictly to the rules published by proper authority.

CHAPTER IX
JUMPING

52. The Seat 123
    a. Effect of shortening the stirrups 124
    b. Common faults from shortening stirrups 124
53. The Horse's Movements 125
    a. How the Horse jumps 125
    b. Head and neck, during approach 125
    c. Head and neck during take off 126
    d. Head and neck during period of suspension 126
    e. Head and neck during landing 127
54. The Horse On the Hand For Jumping 127
    a. Premature jumping 127
    b. Explanation of "On the Hand" 128
55. The Rider's Actions 128
    a. The approach 128
    b. The take-off 131
    c. The period of suspension 132
    d. The landing 133
    e. Summary 134
56. Riding a Course of Jumps 135
57. Refusals 136
58. Run Outs 137
59. Rushers 138

52. THE SEAT.

Any horseman who has had the aesthetic thrill of watching an experienced horse soar over obstacles, when at liberty, has been furnished with a clue as to how a jumper should be ridden. The ease and grace with which he clears formidable fences, when unhampered by a rider, convincingly demonstrates that he requires the greatest amount of freedom the man on his back can possibly give him.

The military seat, without modification, is admirably suited to such jumping as is normally encountered in military work, hunting and cross-country riding. It is also the seat employed in horse show jumping, except for a shortened stirrup.

The shortened stirrup leather gives the greatest mechanical advantage to both horse and rider when large obstacles are encountered. Security is lessened, but it is no longer paramount, as the horse is presumably well trained and jumps freely. The rider depends primarily on balance to maintain his seat, supplemented by the action of the thighs, knees and legs.

a. *Effect of shortening the stirrups.*

When jumping obstacles higher than three and a half feet, the stirrups should be shortened from one to four inches; the higher the jumps are, the shorter the stirrups should be. It will be remembered that this involves the following modifications in the Seat:

(1) The knees are raised and moved the least bit farther forward on the saddle-skirts.
(2) The thighs are raised and consequently form a smaller angle with the horizontal.
(3) The buttocks and trunk are pushed farther to the rear.
(4) The trunk is inclined farther forward from the hips, in order that its center of gravity, which has been moved backward, will remain, as usual, over a point in advance of the center of its base of support. (Since the seat is out of the saddle much of the time when jumping, it will be recalled that the horizontal measurement of the base of support extends from where the inside of the knee is in contact with the saddle to the heel of the boot).
(5) Roughly speaking, the body's center of gravity should be approximately in a vertical plane passing just in rear of the knee joints.

b. *Common faults.*

The most common faults found in beginners, when the stirrups are shortened, are the following:

(1) The knees are too high and pushed too far to the front. As a result the buttocks slide too far forward. In some very bad cases, the knees rest at the front ends of the saddle-skirts, or beyond them.

(2) The heels are high, resulting from the faulty positions of the buttocks and knees.

(3) The stirrup straps are not perpendicular.

(4) The loin and back are humped.

All these faults make the seat weak and insecure. They result from failure to keep: the heels driven far down; the calves closed against the horse; the knees held in against the saddle-skirts; the back and loin muscles contracted sufficiently to keep the back straight and the loin concave. Any one of the above faults entails the others.

53. THE HORSE'S MOVEMENTS.

a. *How the horse jumps.*

Before describing how the seat functions, and the manner of conducting the horse to, over, and beyond the jump, a brief discussion from a mechanical viewpoint of the horse's normal actions when jumping will be given. An animal with a short neck, such as a deer or dog, when jumping a high fence, pops almost straight up into the air and lands on all fours simultaneously. A horse, on the contrary, by using his head and neck as a balancer, describes a graceful parabola, with his forefeet coming to ground well in advance of the hind. *It is an interesting fact that the forefeet strike the ground and leave it again before the hind feet touch.* Inasmuch as this see-saw, or bascule movement is accomplished as a result of the employment of the head and neck, the necessity for the rider's not interfering with the movements of the head and neck during a jump, is evident. All have seen a poor rider hanging onto the reins and pulling the horse's head high in the air over the top of a jump. With his neck thus immobilized, the poor horse also lands on all fours simultaneously and is said to "jump like a deer."

b. *Head and Neck, During Approach.*

In approaching an obstacle, a horse that jumps in good form, lowers and extends his neck in an easy, graceful position in order to estimate the character and height of the obstacle, as well as his point of take-off. Very often he may be seen extending his neck and head far to the front with a very swift gesture just before his front feet leave the ground. This gesture apparently is made for the purpose of producing a counteracting force to the rear, permitting him to check and shorten his last strides in placing himself for his take-off. Also, with an extended neck, he can exert its greatest force in its next and most important gesture. *Thus, during the approach, the rider's hands should be passive and exceedingly elastic, softly following all the forward and backward movements of the horse's head.*

c. Head and neck, during take-off.

Just before raising his forehand, the horse throws his head and neck upward, placing their mass more or less directly over the shoulders. This is an advantageous position, since it is the powerful opening of the previously closed shoulder joints that furnishes the principal muscular force in lifting his forehand. In addition, the momentum given the head and neck by the upward gesture tends to pull the forehand up in the first stage of its ascent.

In this phase, the head and neck are used just as a man uses his arms in making a standing high jump. Here again, the reins in no way should impede the action of the head and neck.

Often they will go slack due to this rearward gesture, and since the jump has begun, it is well if they do. Poor riders bring the hands in to the body at this phase, pulling the mouth and greatly hindering the horse's efforts. They explain it is "lifting the horse." But intelligent horsemen more accurately describe this as "strangling."

d. Head and neck, during period of suspension.

Let the neck be considered as a lever between the shoulders and head. As soon as the forelegs are raised and start passing over the obstacle, the neck and head, (already across), are extended and thrust forcefully downward by a muscular effort of the neck. This exerts a counter force upward at the other end of the lever, which helps lift the shoulders and forelegs over the obstacle. The inertia of the head gives the muscles of the neck a point of support in exerting their force. This, perhaps, accounts for the fact that many excellent jumpers have rather large heads. Obviously, it is very important that the head and neck, in making the downward gesture to help the forelegs over, be hindered in no way by the reins.

Meanwhile, the hind legs having been engaged well forward under the belly, the hocks and other joints extend and propel the horse's entire mass upward and forward. As the hind feet leave the ground, the period of suspension begins. Immediately after the forelegs have cleared the obstacle, the head and neck are abruptly raised from their lowered position. A similar phenomenon to that produced by the downward gesture occurs; the muscles of the neck act against the inertia of the head as they swing it upward, and the counteracting force at the other end of the neck drives the shoulders downward. It is also this force that gives the horse a see-saw motion about his center of gravity, sending his fore feet to the ground well ahead of his hind ones.

Since the horse is entirely in the air at the time the head is thrown upward, the same interior muscular forces which drive the shoulders down, react again through a second lever, the body, and lift the hind quarters up. This second reaction greatly assists his clearing the obstacle with the hind legs, which are also being flexed and tucked up at the same time. *It is most essential that the rider stay out of the saddle at this phase of the jump.* Settling into it as the hind legs are clearing inevitably destroys the effect toward raising the hind quarters produced by the gesture of the neck, and often knocks the hind legs down into the obstacle.

e. Head and neck, during landing.
When landing, several other gestures are made by the head and neck to soften the shock to the forelegs, as well as to raise the forelegs in the quick hop which they make to clear the way for the hind feet which come to earth a fraction of a second later. For a detailed study of the horse's gaits and the effects produced by the head and neck, *Les Allures, Le Cavalier*[5] by L. de Sévy, is highly recommended. M. de Sévy made several other profound studies of the movements and reaction of the horse and rider.

With this mental picture of the horse's movements during a jump, and with a knowledge of the importance of the gestures of head and neck, certain conclusions will be drawn regarding the corresponding actions desirable on the part of the rider.

54. THE HORSE "ON THE HAND FOR JUMPING."
a. Premature jumping.
No attempt should be made to ride a horse over obstacles until he understands and readily obeys the aids. First and foremost, he should accept the bit, or in other words, "go nicely on the hand." Secondly, he should be easily controllable at any speed. Many trainers, unfortunately, start jumping their horses, and even entering them in horse shows, before they can be ridden at a gallop around a definitely prescribed circle. Potentially marvelous jumpers soon become unmanageable and are ruined for the hunting field or the show ring, by this illogical procedure.

---

5   English translation: *Seats, gaits and reactions* by L. De Sévy - Editor's note.

b. Explanation of "On the hand."

When the horse is "on the hand," he necessarily must be "in front of the rider's legs," otherwise he cannot be kept "on the hand." This signifies that he is instantly responsive to the legs, increasing the gait smoothly and gradually at their demand. He carries his head fairly low and the neck well extended, accepting the frank feel of the bit calmly and without pulling. To his rider, he feels thoroughly committed to the forward movement at all times. If the rider increases tension on the reins, and squeezes with the legs at the same time, the horse remains calm while thus more tightly enclosed by the hands and legs. He cannot stop as long as the legs or spurs urge him ahead. The rider has the sensation that the horse's center of gravity is near his shoulders and always in front of the rider's legs. There is a slight preponderance of weight on the forelegs, which causes him to seek support on the bit.[6] At the free gallop, if the horse is a good jumper, the normal feel is a fairly strong one. Remember that the feel increases in intensity as the speed increases.

On the other hand, a horse "back of the bit" and legs, gives his rider exactly the opposite impressions. He seeks to avoid a frank feel on the bit. The rider senses that the horse can stop or slow the gait with ease, and that the aids are almost powerless to prevent it. His center of gravity seems to be back of the rider's legs, and his head raises or is tucked in at any increased tension on the reins in an effort to avoid the bit. At any time, this horse can refuse a jump, or "fade back of the legs and bridle" out of control. It is impossible to ride such an untrained mount over difficult jumps, or any place else, unless through his own good will he chooses to go.

55. THE RIDER'S ACTIONS.

Riding over obstacles may be divided, for the purpose of discussion, into four stages: the approach; the take-off; the period of suspension; the landing.

a. The approach.

The approach may be considered as extending approximately over the last twenty yards prior to the jump. It is in this space that the rider has most

---

[6] Horses, at speed, namely when racing, hunting, or jumping, carry a preponderance of their weight on the forehand, and should take a frank support on the bit. Well-trained horses at slower gaits, keep their weight more equally distributed between the fore and hind legs; consequently they take much less feel from the hand than when going at speed. Highly collected horses carry a preponderance of weight on the hindquarters by engaging the hind legs well under the mass. Therefore, they take an exceedingly light feel on the bit. However, an exceedingly light mouth is greatly different from that of a horse "behind the bit." The former is "in front of the legs" while the latter is "behind the legs."

opportunity to exercise his equestrian skill. He must have determined the proper speed, and rated his horse accordingly. The proper speed will be governed by the size of the obstacle, the nature of the ground in front of it, other obstacles or hazards beyond it, and the speed with which the particular horse jumps best. In addition, certain horses jump much better when galloping with a particular leg leading. In approaching a formidable obstacle, if the time permits, the horse should be put on the lead with which he jumps best. Of these problems, the most important is rating the horse at the proper speed without annoying or exciting him. This takes delicacy and skill with the hands. Briefly, a combination of vibrations and momentary fixings of the hands are generally most successful. Above all, avoid leaning backward out of balance and pulling. About fifteen yards from the obstacle, rating a horse or bothering his mouth in any way should entirely cease. If, up to this point, increased tension and work with the hands has been required to slow the speed, it should from now on be stopped, and the normal feel smoothly and progressively established.

During the last few yards of the approach, the rider must give the greatest attention to following the movements of the horse's head and neck with semi-relaxed shoulders, elbows, and hands. The hands, at this critical period, must not distract the attention of the horse from the jump. *The elbows, particularly, must open and close with elastic smoothness, in following the mouth. Upon arriving at the point of take-off, the tension should never be heavier than the normal feel, and with a trustworthy horse, it is preferable to have the lightest contact possible.*

If the horse is properly in front of the rider's legs, the lighter the tension on the reins, just prior to the take-off, the easier he can be controlled. This is true because the muscles of the neck and jaw are relaxed; there is no pull against the bit to resist. Therefore, the slightest effect with the fingers on the reins will cause the horse to flex momentarily and serve to decrease his speed, keep him straight on his course, or change direction. Keeping the horse straight toward the center of his obstacle during the approach, is very important. Approaching at an angle makes a run-out easy and tempts the horse to try it.

A horseman should always sit down in his saddle during the last fifteen or twenty yards of the approach. If standing in the stirrups, he is unable to feel what the horse is preparing to do, and as a result cannot prevent refusals or run-outs and never knows at what exact stride the horse will take-off. Moreover, with the weight in the stirrups, his legs are not tightly closed against the horse's sides which prevents their acting with sufficient promptness and vigor when occasion demands. As in all else, the legs play a prominent role in jumping.

When seated for the approach, the buttocks should be well to the rear, the loin hollowed-out, the heels driven far down, and the calves and knees glued to the saddle and horse. The body should retain the same marked forward inclination that it has when standing in the stirrups. The tendency to sit up a little straighter should be avoided, since, from the moment of take-off until landing is completed, the rider should be standing in the stirrups.

The good horseman remains quietly in balance during the approach and take-off. Any sliding forward in the saddle, or unnecessary swaying forward and backward, disturbs the horse's equilibrium and distracts his attention at a time when all his faculties are concentrated on placing himself for his take-off.

Normally, at about fifteen yards from the obstacle, after the horse has been rated and the rider is seated in balance, the legs should administer a strong squeeze with the calves, or, in dealing with a timid or unreliable hunter or jumper, a hard pinch with the spurs. The purpose of this leg action is to push the horse momentarily onto his bit more firmly and establish more impulsion from the hind legs by engaging them under him. He is also thus notified that the decision to take the jump has been definitely made, and his only part is immediate execution. After this notification, the calves remain closely clamped to his sides, urging him steadily forward, and ready to act vigorously with the spurs in case he hesitates during the approach. Horses all differ in sensibility to the legs' action, so the rider's tact must regulate their vigor. However, if an error in the amount of vigor is made, be certain that it is on the vigorous side. With a horse prone to refuse, it is advisable to take no chances, but to give him a decisive blow with the spurs when he first sees the obstacle. He then still has time to settle down from its effect and measure his take-off. The legs continue to act moderately all the way to the jump.

Just for a second or two when the legs first act at the beginning of the approach, the hands should resist somewhat if the horse is not strongly on his bit. The instant he accepts it, they begin to follow his mouth. Certain horses depart for the obstacle with a rush. These do not need the leg action at the beginning of the approach. However, if at first they must be restrained, they should be given a lighter feel on the bit progressively, and not by a sudden abandonment of the reins. More important still, the legs must begin to act when nearing the obstacle. There is no horse which will not refuse, sooner or later, if the legs are passive, just prior to, and at, the takeoff. Usually, there is more than normal tension on the reins, due to the necessity of rating most horses when the approach begins. In all these cases, the reins must be fed out progressively, until the normal feel is established. It is a mistake to allow the horse to rush at the jump from a great distance. He usually becomes sprawled-out, or refuses after escaping the bit.

During the last few strides, there is always a moment when it appears that the horse has measured his strides badly and is "in wrong" for the take-off. *At this disturbing moment, the rider must resist the strong inclination to contract his fingers and elbows, while allowing the legs to "go limp."* This reaction to the instinct of self-preservation must be overcome. When the approach appears all wrong, more than ever must the rider relax sufficiently to sit still in the saddle, maintain his forward inclination, de-contract the muscles of arms and hands, follow the horse's head, and increase the squeeze of the legs. In the next fraction of a second, if these directions have been followed, the situation will invariably clear up. The rider's passive hands having left the horse to his own devices, he, urged also by the same instinct of self-preservation, as well as the decisive action of the rider's legs, will place himself correctly and negotiate the jump with ease. The approach demands from the horseman, fast thinking and excellent coordination. He must maintain his balance, seat, and muscular control, without stiffening or standing in his stirrups.

It is noteworthy that most all mistakes at an obstacle are made as a result of the horse getting too close when taking off. "Getting too close" is caused by the restraining influence of apprehensive hands, lack of deciding action by the legs, or lameness or soreness in the horse. Poor hands, which jerk a horse's mouth during the jump, soon make him afraid to jump boldly, since the bigger the jump, the harder the jerk. He creeps closer and closer at each successive jump, and if the punishment continues, ends by refusing.

The horse should be encouraged, during all his training, to take long, free jumps, in his stride. Teaching him to jump in the stride does not mean that the rider should make the serious mistake of over-riding, by violently spurring, flopping his arms, and swaying the body back and forth during the approach. Such wild actions upset the horse and lead to falls. A good horseman endeavors to carry his horse along freely and boldly, with light hands and strong legs, instilling courage and confidence, but avoiding excitement. The legs are always active *on any horse* during the approach, gently encouraging him to "jump big."

b. *The take-off.*

At the take-off and during two or three strides prior to it, the squeeze of the legs should be particularly strong, and the spurs, ready to pinch the horse at any indication of a refusal. The contact with the mouth should have become progressively lighter during the approach, and at the moment of the take-off, should be exceedingly gentle. As the forelegs leave the ground—never before—it is best to allow all contact to vanish, except that maintained by the weight of the reins. The reins may have even a slight sag in them immediately after the forelegs leave the ground, during all the period of suspension, and when landing. It is astounding to see the difficult situations from which a horse can extricate himself during a jump, *if the rider's reins are slack.* Unless the hands are excellent, let the reins go slack after the take-off.

Many of the best hunters and jumpers start their approach for an obstacle with a rush. Upon nearing it, they check their speed suddenly, bunch themselves, and shorten the last two or three strides before taking off. With such horses, the rider must be well forward when the rush starts, to avoid falling back out of balance, which will cause pulling on the reins. Also the sudden checking and abrupt shortening of the stride, commonly referred to as "propping," make it difficult for the rider to maintain his position and balance. His seat tends to slide forward in the saddle, and his lower legs slip to the rear in a helpless position. This not only deprives him of the power to use his legs and of his security of seat, but also disturbs the horse's equilibrium and calculations for the take-off. Moreover, due to losing his position in the saddle, contact with the horse's mouth is suddenly abandoned. The interference with his equilibrium and the sudden loss of support from the rider's hands, often cause the horse to refuse. *It is only by keeping the heels driven far down, the legs and knees tightly against horse and saddle, and the buttocks pushed well to the rear by keeping the loin concave, that the rider can prevent any dislocation of his seat.* If the seat is not deranged, he will not lose contact with the horse's mouth, nor the use of his legs. Accord between himself and the horse will exist, and there will be no refusal. *He will have a sensation of pushing his seat to the rear as the horse checks, while at the same time he can drive the horse forward with his legs.* Keeping the heels down, which allows the feet to brace against the stirrups, while the legs urge the horse on, is the only way that this can be accomplished.

c. *The period of suspension.*
Motion pictures prove that the rider's seat leaves the saddle at the take-off, as a result of the horse's checking and lifting his forehand; not as a result of the extension and propulsion of the hind legs, as is generally believed. Consequently, a forward seat favors the shoulders in their work of lifting the rider's mass at the takeoff. He should, from the moment he is thrust forward and upward, remain out of the saddle during the entire period of

suspension, as well as while landing. This is accomplished by partly stiffening the knee joints and remaining balanced over the knees, stirrups, and heels. He is actually standing in the stirrups with almost all his weight settled in the heels of his boots. If the ankle joints are relaxed, the weight, perforce, drives the heels well down and forces the calves against the horse. The calves and knees by their grip aid greatly in maintaining the position and balance. The angle at the hip closes, and that at the knee opens, as a result of the upward thrust out of the saddle. The back muscles must be immediately tightened, and remain so, while passing over the jump, in order to keep the loin hollowed out and give muscular control of the trunk. This control of the trunk, in a large measure, is what permits the rider to stay in balance.

The important points for the rider to remember during the period of suspension are: first; the horse makes the gesture of extending his neck and lowering his head, to help his forehand over the jump. The hands must therefore feed out all the rein necessary by extending the arms, and in no case, interfere with the gesture by jerking the mouth. Secondly; the instant the forehand has cleared, the head and neck are thrown upward, forcing the forelegs down, and helping to lift the hindquarters over the obstacle. The rider should stay out of the saddle and well forward. Sitting down at this instant will interfere greatly with the hind legs' clearing. On the other hand, remaining forward helps tip the descending forehand downward and aids the horse's efforts.

d. The landing.

This phase begins the instant the hind legs have cleared the obstacle. The horse's forehand is descending. The rider's arms are well extended, to give sufficient rein; his back remains straight and hollowed out, and the brace against the stirrups again is necessary to prevent sliding forward in the saddle as the horse lands. The trunk has been inclined far forward from the hips as the horse rose and cleared the obstacle. As he starts downward, the rider's closed hip joints begin to open and the knee joints to close; he slightly straightens his trunk to move his center of gravity backward and maintain his balance over his knees, as the momentum of the jump dies out. As the descent begins, the horse pivots about his own center of gravity and between the rider's knees, bringing the saddle close to the crotch and buttocks. The rider should not, however, sit down, but remain balanced in the stirrups with the help of the knees. During the entire period, the reins should exert no tension whatever. They cannot assist the horse while he is in the air, whereas the mildest action on the bit may greatly militate against his efforts. The fingers should always remain relaxed and half open during the entire jump.

As the fore feet land, there is, as has been described, another up and down gesture of head and neck in assisting the front feet to hop out of the way of the descending hind feet. During this time, unless it is necessary to immediately change direction, the reins remain lightly stretched or slack. The rider should gently collect them and establish the normal feel as the horse takes his first strides in resuming the gallop. In the meantime, the rider continues to remain out of the saddle, which prevents thumping the horses back as he lands. The ankles, knees, and hip joints may close slightly to soften the jar at this time. The body should not be allowed to flop far forward. This can be avoided by using the back muscles to keep the back straight and the loin hollowed out. After the gallop is resumed and the reins are collected, the rider gently sits down in the saddle if approaching another obstacle. Thoughtless riders are inclined to sit down heavily as the horse lands and snatch the reins taut. The horse interprets the painful effects as punishment associated with jumping, and soon begins to rush, refuse, or bolt. Landing is an ungraceful and difficult phase for horse and rider. The brace on the stirrups, pinch with the knees, and force exerted by the back muscles, to maintain balance, are extreme, after the horse jumps "big." The strain on the horse's forelegs is very great.

Throughout the jump, and especially in landing, the knees pinch the saddle strongly and remain fixed in place. The lower legs, (calves), squeeze the horse until the take-off is actually made, whereupon the rider's weight goes into the stirrups. His weight is voluntarily kept in the stirrups throughout the jump; and while the lower legs feel close to the horse, they move a little due to the weight in the stirrups, and because the stirrup-straps remain approximately vertical during all phases of the jump. It is a common and bad fault to allow the lower legs to swing to the rear, because of riding entirely on the knees and not in the stirrups during the jump. It weakens the seat and leaves the rider very unstable.

e. Summary.
These remarks on jumping may be epitomized as follows:
1. The approach;—body inclined forward from hips, quietly in balance; hands follow mouth; legs active up to and including take-off; leg pressure increases gradually, rein tension decreases; horse directed perpendicularly toward center of obstacle.

2. The take-off;—trunk and seat still; legs active; hands passive; rein tension very light.
3. The period of suspension;—rider out of saddle continuously, in balance in stirrups; back muscles, knees, and hips function in maintaining balance; very light or no tension on reins.
4. The landing;—rider remains standing in stirrups; very light or no tension on reins; normal contact resumed after landing is completed.

56. RIDING A COURSE OF JUMPS.
The rider, after practice over individual jumps, must soon begin to ride a course of jumps. It is there that the difficulties of control and the opportunities to exercise judgment present themselves. Courses should at first be simple and then gradually increased to longer and more difficult courses with many varied obstacles and changes of direction. Here are a few points to remember in riding a course:

a. Before starting the course, review it in your mind so that there is no doubt as to whether you know the course or not.

b. Once you have jumped an obstacle, set your mind on the next one. Don't look back. The next obstacle should have your entire attention.

c. Ride quietly in balance between jumps and take the turns as smoothly as possible. Nothing distracts a horse more than the rider who sits back on the cantle and jerks him around a turn, then suddenly lurches forward just before he takes off.

d. In approaching broad jumps where more speed is necessary, a somewhat stronger feel of the mouth should be maintained during the approach, and the legs should act more vigorously to sustain impulsion in order to ensure that the horse has speed enough to carry him over the obstacle. If in difficulty, the horse should be allowed to "shorten." The rider should not attempt to push him off but should allow him to take the short stride to place himself, while at the same time, maintaining the impulsion and the strong feel on the mouth.

Although before a broad jump a stronger feel is maintained, the hands, through the elasticity of the elbows, must continue to follow the backward and forward movement of the mouth. That is, the hands are never "fixed" so as to resist, nor are the fingers tightly closed. Tightly closed fingers tend to immobilize all joints of the arms and thus prevent following the mouth. In jumping, this resistance of stiff arms is fatal and no horse so abused can possibly become a confident or certain jumper.

e. On straight up and down fences, a very light feel of the mouth as the obstacle is approached permits the horse to place himself. "Over riding" accounts for most faults at straight up and down fences.

f. Oftentimes a rider will arrive at the take-off with a heavier tension on the reins then he usually maintains. This may be the result of apprehension, miscalculation, or the fact that his horse has bolted too fast at the jump. In any case, the fatal error of suddenly abandoning the reins one, two, or three strides in front of the obstacle, must never be made. It may cause a grave accident, since the horse is completely unbalanced by this sudden loss of strong support.

57. REFUSALS.

With a refusing horse, contact with the mouth must be maintained until he actually leaves the ground with his forefeet at the take-off. In order to stop at a jump, if he is going along at a good pace, he must lower his head. To duck his head, he first must escape contact with the hand. *Therefore, with a refuser, contact must never be lost.* At the same time, the rider's legs must act more vigorously than usual, since the feel on the mouth must be heavier to prevent escaping the bit and ducking the head. *It is, however, very necessary that the hands follow the mouth while maintaining the heavier contact.* The usual error is committed through fixing the hands or pulling when keeping the horse's head up, which by immobilizing his head and neck, prevents his placing himself and makes it physically impossible for him to jump. Almost invariably, a refuser ducks his head either to the right or left when stopping. After determining which side he habitually chooses, the opposite rein should have a little greater tension during the last few strides. As usual, after being forced to jump, he should be given his head entirely as he takes off.

Since almost every refuser invariably turns toward a particular side as he ducks his head, the rein on the opposite side should be predominantly tightened during the last few strides in order to block the route of his favorite escape. The other rein's tension is mildly diminished. As a refuser, until he has abandoned the vice, he must be held quite strongly in hand with consequent limited freedom of head and neck; it follows that he cannot as accurately place himself for his take-off. Nevertheless, in all training, one thing must be corrected at a time, and although he undoubtedly will make mistakes at the obstacle, he first must be convinced that refusing is impossible. For the rider's safety, it is best to work such a horse over obstacles that are not too rigid.

The moment after a refusal occurs, the offender should be faced squarely up against the center of the obstacle and punished sharply, just in rear of the cinch with the spurs or on the croup with the riding whip. Punishment is of no value whatever if administered as much as five seconds after the offense is committed. He then should be *turned in the opposite direction from that toward which he ducked* and taken back for another trial. For instance, in leaving the obstacle he should never be allowed to turn to the left if that is the direction in which he turned when refusing.

With refusers, as in all cases where defenses arise, it is best to revert to more elementary work and go over all early training in jumping. Nevertheless the horse must be forced to jump, *at the present time,* the identical obstacle refused, even if it is necessary to lower its height. Also, a thorough diagnosis should be made to discover if pain is the cause of the refusals.

58. RUN-OUTS.

Running out is accomplished by a sudden check or a lunging increase in speed, combined with dropping back of the bit in the first case, or throwing the weight on the rider's hand in the second. In either case, the horse needs more suppling with the direct rein and leg on the side away from the run-out. The opening rein, *or direct rein of opposition*^ is used to prevent run-outs, for if a horse is forced into his jump with the *indirect rein of opposition,* his head is turned away from the obstacle and even though he is unsuccessful in running out, he very probably, because he cannot see well, will commit a serious mistake in jumping. If the rider, with a horse inclined to pop his shoulder out and to run out, either by increasing or by decreasing his speed, vibrates the bit well away from the obstacle, the defense can be broken up before initiated. In increasing the speed as he runs out (to the right, for example) a horse normally takes a strong feel of his bit and throws his weight violently onto his right shoulder. Consequently, if the rider, anticipating this trick, uses vibrations, the horse is unable to set his jaw against the bit and hence is unable to throw his weight on his right shoulder.

In run-outs accomplished by dropping back of the legs, and a decrease in speed, the seat is often displaced as in a refusal. Driving the horse up to his bit with the legs and using vibrations quickly will break up this form of disobedience. The horse needs lessons with the spur to inculcate frank, forward movement.

If a horse succeeds in running out to the left, for example, he should be instantly turned to the right by the right direct rein while being punished simultaneously with the right spur (lateral side). As he becomes docile and obedient to the aids, their action ceases and calmness is restored. Under no circumstances should he be permitted to turn about to the left, after running out in that direction, as the rider conducts him back for another trial at the obstacle. After preliminary vibrations, the right rein predominates during the next approach and both legs are very active.

As in the case of refusers, the most effective and permanent corrective measures lie in first resuming elementary suppling and control exercises with no jumping involved, followed by a rehearsal of all preliminary exercises over obstacles.

59. RUSHERS.

Almost without exception, rushers are produced by heavy hands. The average colt, properly trained and gradually prepared for high obstacles, never acquires this disagreeable trait. It usually is inspired through the trainer's tightening the reins the moment an obstacle is faced, whereupon the horse, realizing that he needs a certain amount of momentum to jump, takes control of matters and bolts forward despite the rider's pulling. Not only does the horse realize he needs some speed, but when the heavy hands fail to allow him liberty of head and neck, he becomes frantic. Always a frantic or frightened horse seeks escape by running away. As mentioned previously, many excellent jumpers, particularly at the first obstacle of a course, gather impulsion quickly. If the rider is unperturbed, stays in balance, and follows the mouth from the first stride with light hands, his horse invariably steadies his gait so as to accurately gauge his take-off. This does not mean he should slow down and "peck" before each jump, but if he has been "let alone at his jumps" during training, experience will have taught him to measure carefully his last three or four strides, in order to take off at the most advantageous spot, just as the human high jumper does.

With all vices, a cure is much more difficult and laborious than is prevention. Work on the longe for horses that refuse, rush, or run out, will accomplish wonders. The rule, heretofore given, concerning exercises to calm and relax the horse prior to any jumping must necessarily be followed. With the rusher, to teach proper acceptance of the bit with the lowered and extended head and neck is the prime essential. It will be noted that most offenders of this type are horses which have been over-flexed, and fold their heads and necks into balls the moment their riders attempt to restrain them before an obstacle. To jump a rushing star-gazer is a dangerous ordeal. If he is rated, he cannot always see the jump, and if turned loose too late, a calamity is always imminent. Until his head carriage is corrected, he should not be jumped.

Quietly galloping on a small circle just opposite one end of a small obstacle until a rusher is calmly on his bit with extended head and neck, and then occasionally swinging wide so as to let him hop over it, helps to cure rushing. Needless to say, the rider should allow an absolutely loose rein and hold to the pommel of the saddle, if necessary, until the horse discovers that his mouth will not be bothered while jumping. The rider must also be sure that his horse has not "gone asleep" under him. He should be aware of the jump ahead. Many bad falls have occurred this way. If the horse is not attentive to the jump ahead, a mild vibration of the reins to attract his attention to the jump should be resorted to. It also aids if, beyond the obstacle, the horse is faced at a few yards by a wall or other insurmountable barrier, so that after jumping, he has "no place to go." After each jump, the rusher should be gently halted and allowed to rest at a walk. Of course when out-of-doors, he should not be jumped over obstacles while going toward his stable.

CHAPTER X
CROSS COUNTRY RIDING

    60. Value to Animals and Their Riders 140
    61. Terrain.. 140
    62. Seat and Hands... 141
    63. Obstacles 141
    64. Slides 142
    65. Riding in a Flock 142
    66. Falls of Horse and Rider 143
    67. Consideration of Horse and Rewards 143

60. VALUE TO ANIMALS AND THEIR RIDERS.
The benefits derived by the horse are improvement in condition, building and developing muscles, attainment of natural balance, development of confidence in his own ability and powers, and quietness.

It develops bold free riders and teaches them the capacities and limitations of the horse. It is essential that the military horseman be able to conduct his mount cross country without undue fatigue and without undue loss of time. The principles of equitation find their application in cross country riding.

61. TERRAIN.
The best country for riding is a varied terrain in which steep hills of moderate height, slopes of varying degrees, rocks, timber, fordable streams, and small valleys are at hand. On flat level ground, horses are more liable to run, as there is nothing to attract their attention or to cause them to change their balance. Hills and slopes, rocks and small ditches cause a horse to extend and collect, to lift his feet and pick his way, to maintain

his balance and equilibrium under the various changes of his center of gravity. In other words, the horse learns to keep his legs under him and stay on his feet. Wooded country serves the same purpose; it makes the horse pick his way and causes him to change direction suddenly and of his own accord. Streams should be available to give the horse practice in fording and swimming.

62. SEAT AND HANDS.

The Military Seat is well adapted to hunting and cross country riding. No changes are necessary.

Normally, the rider should keep a constant contact with the bit by means of a softly stretched rein. He should not pull on the horse's mouth, which the horse resists by throwing his weight on the forehand and increasing the pace. When necessary, half-halts, or the vibrating rein should be used to slow and quiet the horse and lighten the mouth.

Over well-known terrain, with good footing, the horse may be ridden on a loose rein. Loose reins permit the horse complete freedom in the use of his head and neck, which enables him to obtain his natural balance, gives him confidence in himself, induces calmness and therefore helps to teach him to go quietly. Therefore, the horse should sometimes be ridden cross country in this manner. However, in rough, doubtful, obscure footing the horse should be ridden with a soft or firm following hand to guide and steady him in his passage, followed by the reward of a pat on the neck and words of praise after the negotiation of this hard going. Riding with loose reins should not degenerate into license for the horse to go any place or at any gait he chooses. The rider must impose his will upon the horse by use of legs, weight, voice, and short but frequent use of the rein aids.

63. OBSTACLES.

In general, obstacles should be approached squarely in a calm, confident manner with a firm determination in the rider's mind to take them. The horse has less tendency to refuse or run out if he is brought up to the obstacle at a right angle. Also, if riding in company with other horsemen, by riding squarely upon the obstacle, there is no danger of fouling. If the rider shows apprehension either by his hesitancy in the approach or by a sudden and rough use of the reins, legs, or whip, it is transmitted to his mount, whose performance is naturally unfavorably affected.

Obstacles of height should be taken at a gallop of moderate speed with the horse going freely, calmly, and confidently.

Obstacles of width should be taken with the horse forced up strongly onto the hand so that he cannot decrease his pace upon approaching the obstacle.

64. SLIDES.

In taking a slide the horse should be kept well under control by use of the reins and legs which make him start down the slide straight, and prevent any attempt to turn aside at the instant when his fore feet have started down the slide but the hind feet have not. The reins are normally held in both hands, but they may be held in one hand if the other is required for the use of weapons or other purposes. The rider's legs should remain closed against the horse, and his body inclined forward and kept there. This position allows the rider to go down the slide with his horse without danger of straining the jockey muscles. If the body is inclined backward when the horse slides, the legs are either pulled forward and up from the sides of the horse with the danger of losing the seat, or a terrific strain is put on the muscles of the legs and thighs, which is liable to strain the jockey muscles. The forward position also frees the horse's loins and allows free play of the arched muscles in the loins, which come into play when the horse brings his hocks up under him in sliding. There is less shock to the horse at the bottom of the slide if the rider is forward, because the shoulders are attached to muscles which can absorb the shock easier than the bony structure of the hindquarters.

When the footing is bad, such as mud, ice, soft or rocky terrain, hard roads, the horse should be kept going straight ahead at an even pace rather than being slowed up. So long as he continues on straight and free (no chugging up), there is little danger of slipping and falling. The same is true of jumping on slippery terrain.

65. RIDING IN A FLOCK.

When riding in a flock, it is best to start with slow gaits and gradually increase the gait as the horses warm up.

The ground should be chosen for the horse, and suitable gaits selected for work over rough or difficult terrain. The riders place themselves within the flock wherever they desire, the only limitation being that individuals must not permit their horses to go out in front of the person conducting the ride. This precaution is taken so that the leader can maintain an even gait and select the ground on which he desires to change gaits.

The principal reasons that horses pull when riding in a flock are, besides the lack of tact on the part of the rider, fast gaits when going in the direction of stables, or fast gaits down long slopes or into the wind.

Riders should be careful not to follow in trace behind other horses, for in case of a fall or a refusal at a jump, this will often result in a serious accident. Each rider must ride his own line.

The ability to maintain a uniform gait while riding over fairly rough ground or up and down gentle slopes is a good test of the rider's tact and skill in the use of the aids. It calls for continual fine and well judged applications of the aids to control and support the horse and keep him going evenly.

66. FALLS OF HORSE AND RIDER.

If the horse stumbles and starts down, he should be allowed to have his head. It is easier for him to regain his balance and stay on his feet if he has the free use of his head and neck. If the horse does go down, it is better for the rider to turn loose, fall relaxed, and roll away from the horse, rather than to try to stay on while the horse is going down. It is best to go off on the side toward which the horse is falling and roll away from the horse's body.

The rider's momentum is in this direction and a push with the arms or legs at the instant of leaving the horse helps to clear him, and there is not so much danger of being struck by the animal's feet when he struggles to regain his footing.

67. CONSIDERATION OF HORSE AND REWARDS.

In riding cross country, pick the ground which you ride over. If by going to the right or left a short distance you can avoid a bad pile of rocks, stream crossing, or anything that involves extra effort for your mount, do so, unless the passage of difficult terrain or the taking of an obstacle is part of the training of the horse or rider.

Let your mount rest when he is tired. Do not ride for long periods at a fast gait. Most horses will go and continue to go until they are completely exhausted. A little rest in time may save a strained tendon or your animal's wind. After a satisfactory passage of a difficult obstacle, reward your mount by a pat on the neck and a kind word. This helps the understanding between mount and rider.

CHAPTER XI

SUGGESTIONS FOR INSTRUCTION

> 68. General 144
> 69. Qualities of the Instructor 144

68. GENERAL.
a. The object of all training in equitation is to produce horsemen.

b. In order to become a good horseman, it is essential that the beginner receive the necessary instruction and practice to enable him:
(1) To ride with facility either a well trained horse or a difficult, high spirited one;
(2) To break and train a green horse or reclaim a spoiled one;
(3) To condition and care for horses;
(4) To judge the quality, suitability, and capability of horses; and
(5) To have a knowledge of the care and use of essential articles of horse equipment.

69. QUALITIES OF THE INSTRUCTOR.
a. In order that the instructor may produce good horsemen, it is obvious that he must be a skilled and experienced horseman himself. He should be efficient and practical in stable management and in all that pertains to the care of horses. His own stables, equipment, and horses should set a high standard in this respect. He should be a competent judge of horses in general.

b. In voice and manner the instructor should be courteous and considerate, yet deliberate and firm in all his decisions and orders. His general bearing, as well as his dress and his deportment, should be faultless. He should possess special aptitude for the handling of men. He should cultivate a certain facility and address in imparting instruction. His explanations should be as short as possible consistent with completeness and clearness. His ideas and reasoning should be presented in logical order, and should be based upon facts which are obvious and natural and therefore easily understood.

The instructor should constantly display his fondness for horses. He should, by his example and leadership, develop in each beginner not only a spirit of boldness and enterprise on horseback, but also a habit of intelligent solicitude for the health, condition, and continuous serviceability of his mount.

c. Throughout the training the instructor should exercise constant supervision with the aim of insuring steady progress for both rider and horse. He should first clearly comprehend the capabilities and limitations of rider and horse, and thereafter direct the training so as to call forth the maximum efficient effort of both without, however, hazarding the future serviceability of the horse. He should encourage the timid rider to increased effort, and restrain the more courageous one from excesses.

d. In his training of both the horse and the rider, the instructor should apply the same general principles. He should make the instruction progressive in character; he should look to the gradual upbuilding of health and strength in accordance with a carefully arranged schedule. Instruction should be suited to the capacity of beginners. The watchfulness of the instructor and the system and thoroughness with which he proceeds from one step to another in the course of instruction should serve to exempt the beginner from falls and other accidents. Great soreness and fatigue and numerous injuries and abrasions among beginners are indications of a lack of judgment on the part of the instructor. If the more difficult work is entered upon too soon, the confidence of the beginner may be quickly destroyed and his general attitude toward horsemanship adversely affected.

e. There should be frequent rests. During the longer rest periods, the instructor should discuss some subject that is interesting to horsemen, such as the comparative age, height, weight, conformation, gaits, and peculiarities of certain horses. He should emphasize always the good points and good qualities of the horses discussed. He should also point out any unsoundness and blemishes, encourage the riders to examine them, and explain informally the causes and effects of such conditions.

f. Impress upon the pupil early in his training that a fall will not hurt him if he will fall away from his horse. He should under no circumstances hang on to the reins or attempt to cling around the horse's neck. This cannot be too strongly emphasized, as most bad falls are caused by the rider not giving the horse the necessary freedom to get himself clear.

g. Jumping is an important part of the training of the young rider and should not be looked upon merely as a sport. It teaches the rider balance, security of seat, and control of his horse at speed. It concentrates in a very short space of time the many difficulties that arise in riding a horse over difficult terrain and permits the instructor to give the rider more detailed constructive criticism.

A person unused to riding is frequently filled with apprehension at the new and unaccustomed motion and at the difficulty in controlling his horse, especially at the faster gaits. This apprehension is not fear in the ordinary sense; it is merely a physical and mental reaction due to a feeling of helplessness. Under its influence, the mind cannot entirely control the nerves, which, in turn, react unfavorably upon the muscular system. The result is that the beginner is anxious and overstrained mentally; his muscles become contracted so that he cannot relax them; his body is more or less rigid and unresponsive to his will. The instructor, therefore, should give attention to combating these nervous tendencies, endeavoring to leave in the mind of the beginner at the close of each day a sense of having successfully overcome each difficulty encountered and of having really enjoyed riding.

h. Fix in mind, before beginning work, a definite program of exercise for the day, and be sure that these exercises are in proper relation to the work of previous days.

i. Be satisfied with a little progress each day and be careful not to demand too much during any given period. Calmness, good humor, sureness of action, and manifest satisfaction with the progress made, preserve the student's tranquility and gayety of spirit, and increase the interest and zest with which they assimilate instruction.

j. Bear in mind that you should never belittle or ridicule either the student's or the horse's ability. It destroys confidence and respect.

k. In the first part of the instruction, the instructor must keep three objectives in mind, for, until the student becomes proficient in these fundamentals, his further instruction will not produce the desired results. These objectives are:

To give confidence to the rider; to give him the proper mounted position, and to lead him to acquire independence in the use of his aids.

HORSEMASTERSHIP PART II: THE EDUCATION OF THE HORSE

CONTENTS: PART II

XII. Preliminary Considerations 151
XIII. Breaking 158
XIV. Elementary Training 173
XV. Training the Horse to Jump 195
XVI. Training the Race Horse 209

CHAPTER XII
PRELIMINARY CONSIDERATIONS

70. Officer in Charge of Training..................................151
71. Cautions to the Instructor 151
72. Care during Early Training of Remounts 154
73. The Object of Breaking and Training..........................155
74. The Mental Faculties of the Horse..............................157
75. Attention to the Fitting of Equipment.......................157

70. OFFICER IN CHARGE OF TRAINING.

The officer detailed for this work should be selected from among those who have already had experience and possess special aptitude. It must be borne in mind that among the first requirements are *patience, common sense* and a *methodical disposition,* without which the most brilliant qualities will fail to produce good results, and may even prove harmful.

71. CAUTIONS TO THE INSTRUCTOR.

a. Progress should be suited to the age and condition of the remounts. It should constantly be borne in mind during the whole course of a young horse's education that one must be satisfied with a little progress each day; demand that, but no more. Always proceed from the known to the unknown and from the simple to the complex. Never ask anything different of a young horse while he is still under the impression of a preceding requirement. Never attempt to combat two resistances at the same time.

b. There should be frequent rests. Periods of work should be positive and energetic, after which the horse should be allowed, in the fullest and most generous manner, to relax completely and to rest. As training progresses, the distinction should be sharply drawn between standing or moving *at ease or route order,* and standing or moving in a state of *attention.*

Rests should be of a character not to produce soreness or stiffness in the muscles of the remounts or to cause chill of the body by exposure to drafts or rain. If horses are very tired or hot, or if the weather conditions are unfavorable, it is better during rests to lead them about.

During rest periods each rider should habitually examine the feet and legs of his horse and should make such readjustments of the saddle and equipment as are necessary.

c. The training of remounts should be conducted without hurry. Intensity and nervousness upon the part of the instructor brings about nervousness and awkwardness in the actions and efforts of the men. Such nervousness and hurry reacts most unfavorably upon the manners and dispositions of horses, causing them to lose their composure and to become, in turn, nervous and difficult to manage. On the other hand, calmness, good humor, dignity, sureness of action, and short ness of speech, coupled with the manifest satisfaction of the officer in charge with the progress made, produce a most favorable effect. To men and horses alike it acts as a tonic; it preserves, so to speak, their tranquility and gayety of spirit, and increases the interest and zest with which the work is done.

The best test of the ability of the officer in charge of the training of remounts is the condition of the horses at the end of their training, the cleanness of their limbs, and their temper.

d. There are many explanations and illustrations that may best be given on foot. Accordingly, in the riding hall, the instructor should, on occasion, dismount.

e. Throughout a course of training for remounts, the instructor and the riders should bear in mind the following precepts:

(1) *Be systematic.*

Before beginning work, fix in the mind a definite program of exercises for the day. Be sure that the exercises for the day are in proper relation to the work of previous days.

(2) *Be patient.*

Do not destroy the tranquility of horses by demanding a performance that is too difficult, or by demanding it too early in training.

(3) *Be tactful and resourceful.*

Take advantage of the most favorable conditions for teaching a horse a new lesson. Never try to train a fresh horse. Undertake nothing new when the horse is excited or frightened. Do not try to train the horse when his attention is distracted. Do not give a new lesson to a resisting horse. Do not send the horse to the stable in the midst of resistances or with a lesson incomplete. Finish the lesson first and then send the horse away calm and tractable.

(4) *Be moderate.*

Begin with the simplest movements and exercises. These understood, proceed to the next, less simple. In the early training introduce nothing complex or difficult. Use continuously the same means to bring about the same results, thus aiding the horse's memory. Ask little but ask it often; it is by repetition that a horse progresses. Nevertheless, do not let a horse continuously execute a movement incorrectly or in a dull, lifeless manner. Demand attention, correctness, and carriage and action gradually increasing in style and manner, then allow a few moments of complete relaxation. Never strain the attention or tax the strength of the horse. Require no position, attitude, or movement which in itself causes the horse apprehension, discomfort, or pain.

(5) *Be observant.*

Do not attribute every resistance or failure of the horse to inattention or stubbornness. These are often due to ill-fitting bits or saddlery, to a poor rider, to lack of condition or approaching unsoundness, to noises, unaccustomed surroundings, or even to the weather.

(6) *Be exacting.*

Do not be content with the simple tracing of the riding-hall exercises and figures. Every such exercise or riding-hall figure has for its object to teach the horse obedience to the aids and to know how to handle himself in doing so. Accordingly, before taking the first step of a movement, the horse should be placed in a position which favors the simple and natural execution of the movement. The movement will then be executed more easily and correctly.

(7) *Be logical.*

Do not confuse the means by which an end is obtained with the end itself. Practically all of the exercises and riding-hall figures are the means by which the horse is rendered easy to manage during ordinary riding. Accordingly, do not use riding-hall exercises as a proof of training or routine drill movements as a means of training. The first are the means by which the horse is trained. The second constitute the test and the proof of training.

(8) *Be liberal.*

Permit the riders to ride the greater part of the time at will, or, if on the track, without regard to distances. They then have a greater opportunity to really control and to correct the attitudes, positions, and movements of their horses. It also permits the horses to assume their individual natural gaits and avoids irritation by forcing them too soon to take regulation gaits. As training progresses, they should, in periodic tests, be able to ride more and more accurately with *fixed distances* on the track, *by threes* in the hall, or in military formations out-of-doors.

(9) *Be tenacious.*

Never provoke a struggle which can properly be avoided. If, however, a serious resistance is encountered, the rider must not evade the issue; he must emerge from the contest in entire control of the situation.

(10) *Be consistent.*

Throughout the course of training, keep in mind the fundamental requirements of ordinary marches, maneuvers, and of actual field service. Rigidly exclude from the training program every exercise that does not bear directly upon the proper preparation of the horses for such duty. At the same time, take care to include in the program every exercise that renders the animals more easily managed or that actually increases their strength and powers of endurance.

f. Whatever may be the value of the instructor or of his methods, the rider's weight and conformation, his fixity or insecurity of seat, his strength or the lack of it, his suppleness or stiffness, his energy or indolence, his intelligence, esprit, and patience—or, on the other hand, his apprehension, nervousness, or brutality are the factors which have great influence on results obtained in equitation. One should consider these things carefully in assigning riders to horses, especially to young horses.

72. CARE DURING EARLY TRAINING OF REMOUNTS.
a. Husbandry of the Remounts.
The officer in charge of the training of remounts should pay scrupulous attention to the cleanliness of the stables, stalls and mangers, and to the general sanitation of the immediate surroundings. Remounts are very susceptible to colds, fevers, and other ailments which, if not guarded against, will seriously interrupt the training. He should, in addition, by daily inspections, keep himself thoroughly informed as to the kind and quality of forage, its preparation, the amount fed, and the frequency of feeding. The horses being young and subjected, moreover, to new and unusual sensations and experiences, accompanied at times by considerable fatigue, sometimes go "off their feed." Their appetites must then be tempted by such substitutions in the forage ration as may be available and by grazing.

b. The physical fitness of remounts.

At the commencement of training, remounts are not usually in good physical condition. They are not accustomed to hard work and their muscles, tendons, joints, and often even their vital organs are weak. Accordingly, their bodies are susceptible to painful injuries if the work is beyond their reasonable physical powers. Therefore, in the training, the work imposed upon remounts should tend to be long in point of time, daily, but very mild in character. It is out-of-doors, in the open air and sunshine, that young horses are conditioned. Long periods at the walk, very short periods at the trot, and shorter and less frequent periods at the gallop, expand the lungs, strengthen the heart, and build up the muscles of young horses. Their training, in exact obedience to the aids, and in jumping obstacles, should be delayed until gentling and conditioning are well under way. Thus managed, remounts remain more tranquil while undergoing training; they are also much less liable to injuries, especially in the region of the legs.

A too-hurried training is hard on joints and tendons; horses are often thus rendered unserviceable before their training is completed. The same adverse results are frequently brought about by sudden and excessive requirements. Accordingly, moderation as to the standards required in the training of remounts and systematic methods for the general attainment of those standards are for the best interests, not only of the mounted service as to efficiency, but also of the Government as to economy in the procurement of animals. For, once a remount reaches, without injury, a fair state of training, he will, thereafter, through his increased strength and hardiness and his greater handiness and balance, render much more satisfactory service as a military horse. He will, moreover, remain sound and serviceable and of use to the Government for a much longer period of time.

73. THE OBJECT OF BREAKING AND TRAINING.

a. General

The education of the young horse should continue for two years, and this rule should not be disregarded, except in case of mobilization. The preparation for his career comprises two phases; breaking, and training, each corresponding to a definite and distinct objective.

(1) *Breaking.*

The objectives of breaking are twofold. First, to physically develop and condition the young horse. Second, to bring him to the degree of training where he can be mounted remaining quiet and relaxed; be ridden at the walk, trot, and gallop; be turned to the right and left, and be halted without abruptness.

(2) *Training.*

The object sought in the training of remounts is to place the horses in a state of condition and training essential to their military use.

A good military horse is one that is sound and of good conformation; in good condition for hard work and to withstand exposure; tranquil; pleasant to ride; and especially it should be perfectly quiet to mount; should move freely and true on a straight line; be manageable at all gaits and in all changes of direction. It should jump or pass obstacles of all nature; endure the pressure of the ranks and leave them freely; bear all parts of the equipment; not be frightened at the sight or sound of objects out-of-doors; be accustomed to the use of arms; be unafraid of water and a good swimmer.

In the training of remounts, great attention should be paid, *first, to their conditioning; second, to their tranquility; third, to their training,* properly speaking. *Any system of training that neglects the conditioning or which destroys the tranquility of horses, is defective.*

b. These two divisions in spite of their special denomination do not constitute two clearly divided periods; they represent together the necessary time for the remount to respond physiologically to the requirements of military service. The words, breaking and training, nevertheless, carry their own ideas which constantly remind the instructors of the great difference in the work that an immature colt can endure and the requirements that may be made upon a horse of six years. One should submit the young horse to the necessarily severe gymnastics of schooling only when his morale on the one hand, and his physical development on the other, allow him to undergo it without fatigue.

It is difficult to state the period of time that is necessary for the proper breaking of remounts. Whatever may be the rider's skill, the physical development of a horse is subject to the laws of nature. Increased work cannot hasten the natural evolution and substitute itself for the effect of time. For colts four years of age, it may be stated that a year is essential for their proper conditioning; for a horse five years of age, six months should be sufficient; for a horse six years of age, four months; and for a seven-year-old or older, two months should be ample.

In training or schooling, the progression is similar to that employed in the instruction of the rider. This methodical order, naturally, proceeds from the simple to the complicated; it regulates the demands of the rider according to the ease with which the horse is able to answer them.

Care must be taken, particularly in the beginning, to execute the movements under the same conditions and in the same manner until the horse is confirmed in his knowledge of the rider's demands by the effect of repetition. It is only gradually that an obedience, at first laborious and uncertain, will be later transformed into almost instinctive habit.

It is assumed that in the above cases the horses are sound, and of good quality, bone, substance and conformation, and in good health. If such is not so, a longer period of time in each instance may be necessary. In all cases, however, it must be distinctly borne in mind that the assignment of remounts to an organization does not mean that their training is complete. Their training is never finished, but is carried on continuously throughout their service.

74. THE MENTAL FACULTIES OF THE HORSE.

In order to train the horse, his mental as well as physical aptitudes must be taken into account. Therefore, in training, advantage must be taken of those characteristics most frequently manifested by the horse. These mental characteristics were discussed in Chapter II and they should be carefully studied before starting the training of a remount.

75. ATTENTION TO THE FITTING OF EQUIPMENT.

There is. nothing more injurious to the temper of a horse, or that more interferes with the achievement of satisfactory results in his training, than ill-fitting and uncomfortable bitting and saddlery. Accordingly, from the beginning to the end of the course of training, the proper fitting of bits and saddles should receive systematic and continuous attention.

CHAPTER XIII

BREAKING

76. Object 158
77. Importance of Work 159
78. Leading 159
79. Longeing 159
80. Accustoming to the Saddle 163
81. Mounting Lesson...164
82. Mounted Work...165
83. Elementary Lessons in the Aids....................165
84. To Move Forward..166
85. Halting 167
86. The Turn..167
87. First Lessons at the Gallop............................168
88. Preliminary Conditioning..............................169
89. Value of the Various Gaits.............................171
90. Defenses of the Young Horse........................172

76. OBJECT.

The objectives of breaking are two-fold. First, to physically develop and condition the young horse. Second, to bring him to the degree of training where he can be mounted remaining quiet and relaxed; be ridden at the walk, trot and gallop; be turned to the right and left, and be halted without abruptness.

Its principal objective, then, is to progressively condition the remount. The various steps through which the remount has passed, including the time spent in the remount depot, where he has undergone a certain amount of work, serve as the beginning of this conditioning and facilitate breaking.

77. IMPORTANCE OF WORK.

Work is the most important factor in breaking. Besides the role which it plays in the development of the organs, of the young horse, it is the means by which his health is regulated and character developed.

If the young horse does not work enough, he becomes too fat, too playful; he blemishes himself under his own weight, increased by that of the rider, and he spoils his mouth by struggling against the hand that seeks to hold him down.

Nevertheless, the remount should be in rather high condition.

He should have long slow rides out-of-doors, and shorter ones in the riding hall.

Felt or flannel boots may be used for the protection of the legs, especially during work on the longe.

78. LEADING.

Leading at the side of old horses during the first few days is an excellent exercise, permitting the remount to expend his energy without danger to his legs, to become accustomed to outside objects, and to become calm, which is indispensable to his progress. Numerous occasions when it is necessary to lead military horses, make this a useful lesson, though it is not necessary to keep it up very long. In these walks, the remounts should be led first on one hand, then on the other, to avoid always bending the neck to the same side.

79. LONGEING.

The *cavesson* and *longe* and their adjustment have been described in Chapter III.

In lieu of the cavesson, the halter or snaffle bridle may be used as a serviceable, though much less satisfactory, substitute. The halter cannot be adjusted to prevent its slipping around the remount's head. The longe, used with the snaffle bridle, passes through the near snaffle ring and attaches to the outside ring. This is rather severe, and may be painful to the mouth. Moreover, since the longe attaches behind the nose instead of in front of it, as with the cavesson, there is considerable loss of leverage. Satisfactory results can be obtained, however, even with these handicaps.

The *object of longeing* remounts is fourfold; to exercise, discipline, gentle, and supple. The great leverage over the forehand, afforded by the attachment of the longe to the ring on the fore part of the cavesson noseband, gives the trainer a powerful and painless means of control which favors disciplining or obtaining the mastery over the remount, and gentling or gaining his confidence.

The *theory of longeing is based upon the idea of* driving or urging forward with one, the free hand, and restraining or controlling the movement with the other, the longe hand. When longeing to the left, for example, the trainer faces the long axis of the horse, which naturally requires him to hold the longe in the left hand with the bight of longe and driving whip (if used) in the right. The bight of the longe is looped or folded into figures of eights (not coiled) so that it may play out freely as required. The loops also serve to reinforce the driving effect of the hand. The driving whip when carried is held butt to front, hand low, in order to avoid unduly exciting the remount. The hand is swung around bringing the lash end forward when required; or if considerable urging is necessary, the whip may be held naturally, lash end forward.

Longeing should be done equally on both hands, unless the animal through some stiffness or other resistance, objects to a particular side, in which case more is done on that side to overcome the resistance and balance him. Frequent changes of hand should be made, especially in the beginning, with lavish reward for each good performance. Later, when the animal longes well and can be worked on a long longe, such frequent changes of hand will not be necessary or desirable.

The voice is a very important factor in longeing. Horses react almost miraculously to the tone of the voice. The trainer must be very careful to modulate his voice according to the effect desired. A sharp, high tone stimulates the horse. A harsh tone excites him, while a calm, low and prolonged tone quiets and soothes him. When it is desired to pass from the halt to the walk for an example; the trainer should repeatedly say *"walk"* in a firm and somewhat sharp tone, or in moving from the walk to the trot he should say *"trot"* in a similar tone. In decreasing the gait, the voice should be lowered and the words drawn out in their enunciation. Obedience to the voice on the longe is very important as it is a very useful and necessary aid later in the mounted lessons.

Successful longeing depends upon the tact of the trainer. Every movement and every sound made by him has its effect upon the horse. The raising of a hand, a hurried step, being out of position, failure to anticipate the horse's action, use of the wrong tone, too much or too little drive, any of these may result in the loss of control. One of the most important, if not the most important factor in longeing, is the maintenance by the trainer of the most advantageous, or what might be called "the strategic" position, with respect to the horse. This position is opposite or slightly in rear of the point of the hip, and a suitable distance away from him, depending upon the length of longe used and the state of training of the animal. In this position, the trainer is enabled to furnish the necessary amount of forward drive, and at the same time the amount of traction on the longe needed to restrain and prevent escape outward from the circle. When lost it should be recovered quickly but without abrupt movement. The skillful trainer does not lose it. He watches every movement of the horse, anticipates his intentions, and by constantly maneuvering keeps himself always in position. During the first lessons in longeing, the trainer, in his maneuvering for position, usually gets more exercise than does the horse. The condition soon passes, however, and as training progresses the trainer moves less and less, until finally he moves only sufficiently to keep himself in "the strategic" position. The horse should not be allowed to lug on the hand. By giving and taking on the longe, easing the hand as the horse gives inward, he is taught to go on a lightly stretched longe. This prepares the horse for mounted work on a light stretched rein. He should not be allowed to turn inward and cause the longe to go slack. This can be prevented by an appropriate flip of the longe hand sending a slap of the longe out to the ring on the noseband. In the case of remounts, care must be taken not to upset them with these actions, and as the remount turns inward, the trainer will probably have to move in towards his hindquarters to maintain his relative position until the remount has gained confidence enough to respond to the above means of keeping him on the circle. Throughout the lesson the horse should be kept out true on the circle. In each lesson the horse is started off on a short longe which is gradually played out to the desired length. The pace or gait is increased by command progressively from the halt up to that desired, and decreased similarly down to the halt, all on the circle. If the horse unavoidably gets out of control and faces inward, turns outward or reverses on the circle, he is halted, put back on the circle in the original direction and the exercise resumed. Whenever the trainer approaches the horse he should keep the longe stretched by taking up and re-looping the slack.

Should the trainer lose control and be unable to reduce the speed or halt the remount, he may, while continuing the use of the voice, resort to abrupt pulls on the longe rearward, or sharp snaps of the longe against the ring on the noseband, or else, by moving in toward the wall, force the remount to halt.

Longeing is greatly facilitated by having a surface to longe against. The wall of the riding hall, a fence, the side of a building, any sort of barrier which will prevent the animal's escape in that direction, gives the trainer considerable advantage. A corner formed by the junction of two such barriers is even more advantageous, while a complete enclosure, forming a longeing pen, affords the maximum assistance. Much time may be saved by having a few longeing pens either square or circular and some 10 to 12 yards in diameter, in which to give the first lessons to the more obdurate individuals.

To start the remount longeing, the trainer holds the longe with his right hand about 18 inches from the horse's head; the remainder of the longe, folded in figures of eights, is left in the left hand. After giving the remount confidence by petting, the trainer moves forward, pulling slightly on the longe, at the same time clucking to him. He moves thus, around the riding hall, on straight lines and gentle curves, gradually making more abrupt changes of direction, and stopping frequently while saying "Whoa." At each halt he pats the horse. This lesson is repeated on the other hand. When the remount moves forward willingly at the cluck of the tongue, and halts at command, he is ready to start longeing. The trainer, on the near side, takes the longe in his left hand and the slack in his right. He starts the remount moving forward and goes with him on a small circle of 4-6 meters in diameter to the left. Having given him the idea of traveling on the circle, the trainer passes from the operation of *leading* to that of *driving*, by quietly slipping rearward towards the strategic position, at the same time clucking and urging forward gently with the right hand. As he goes rearward to position, he plays out sufficient longe, and stays in fairly close to the remount's side to prevent his turning in and stopping. This is a crucial moment, the transition from *leading* to *driving*, but not a difficult one if the trainer is tactful in gaining and maintaining his position, and uses just the right amount of drive and no more. When the remount has taken a few steps correctly, he is halted and rewarded to show him that he has done well. Then the exercise is repeated, and the remount allowed to go once or twice around the circle before being halted and rewarded again. Now the trainer passes to the off side and repeats the lesson to the right. As the remount becomes more responsive and obedient the longe is gradually lengthened. When he has learned his lesson at the walk, the trot is taken up, and frequent changes made from the halt to the walk, the walk to the trot, and reverse. Since the trainer has much greater control with a short longe, he should always return to it for a new step, or in case of any difficulty encountered. The gallop is not attempted until the remount is well schooled in longeing, and then on a fairly long longe. On account of the speed and exciting nature of the gait, the remount is apt to play up, hence to gallop him on too small a circle would be risking the danger of injury from a slip or fall.

Whenever, in the course of training, a horse develops a serious resistance and becomes difficult to manage, he should be put back on the longe and given a stiff work out. This will serve to tire and relax him, and to reestablish discipline.

After a horse longes equally well to either hand, it is important that the work be continued by constantly varying the size of the circles. Work on small circles balances the horse and supples him throughout; it also tends to slightly collect him by causing him to engage his hindquarters more fully; while work on a large circle permits him to extend freely and adds to his impulsion.

80. ACCUSTOMING TO THE SADDLE.

When the horse has been calmed by work and is perfectly gentle on the longe, the trainer should profit by this condition to gradually induce him to bear the weight of the saddle, and the pressure of the girth. This lesson might prove difficult if attempted prior to longeing.

An assistant should assist each rider the first time he saddles his remount. The assistant, taking the position of *Stand to horse, in front*, attracts the attention of the horse by his firm but gentle manner and the soothing tones of his voice. The rider, meantime, with his saddle on the right arm, approaches the horse from the left front, runs his left hand along the horse's crest and withers, strokes him on the shoulder and back, and places the saddle quietly in its place. The girth should be adjusted comfortably; it should not be drawn tight. The stirrups, in a difficult case, are previously removed from the saddle; otherwise, they are run up and secured so as not to swing about and frighten the horse.

The horse is again worked on the longe, at the walk and trot, to accustom him to the feel of the saddle. During the first lesson he should not be worked at the gallop; at that gait he is the more easily tempted to resist, a thing that should be strictly avoided.

The horse is halted frequently and petted, his legs rubbed, his feet picked up, and the saddle handled and pressed upon. Eventually the girth is tightened as it should be normally, the stirrups are let down, and work on the longe is resumed.

He is thus prepared for the mounting lesson, which becomes an easy operation. It is a fact that defenses nearly aways result when young horses are saddled and mounted for the first time, on the same day.

81. MOUNTING LESSON.

The instructor determines according to circumstances, the most opportune moment for giving the mounting lesson, but he should profit by the fatigued condition of the remount at the end of a work period, at which time the remount is most apt to be calm.

The lesson should be given during the work on the longe, but only after the horse has been relaxed by the exercise of longeing. He personally directs the first mounting lesson, which is given to each horse individually, and displays the greatest calmness and patience in this lesson.

Accompanied by an assistant carrying a can of oats if need be, the instructor places himself squarely in front of the horse and pats him; he takes hold of the longe in his left hand 18 inches from the noseband with the looped end in his right hand. The longe should be vibrated when necessary in such a manner as to keep the attention of the remount. The rider approaches the horse's head, pats him on the forehead, over the eyes, on the neck and haunches. He slaps the saddle, lowers and raises the stirrups. He mounts smoothly without hesitation or hurry. If, during the lesson, the horse moves out of place or breaks away, the instructor leads him forward and quietly begins again.

When putting his foot in the stirrup the rider is careful to point his toe downward so as to avoid touching the horse's side, which might disturb him. He should not pause after raising himself in the stirrup, for, when his entire weight is in the stirrup on one side, it breaks the equilibrium and makes his task more difficult. Prior to mounting the right stirrup strap should be twisted so that the stirrup hangs perpendicular to the side of the horse. This facilitates the placing of the foot in the stirrup. At times it is advisable to use the right hand to assist the right foot into the stirrup.

Generally, the horse should not be moved forward as soon as the rider is in the saddle, so as to prevent the idea of the forward movement from being associated with that of reception of the rider's weight.

It is best for the first few times to terminate the day's lesson by the mounting lesson, so that the cessation of work and his return to the stable may serve as a reward.

Those horses that are found very difficult are immediately put back on the longe.

The first lesson in the forward movement with the rider mounted should follow only after the horse submits quietly to mounting and dismounting. The instructor should move the horse forward on the longe. The rider, without applying aids, remains quietly in the saddle.

Mounting lessons should be given on both the near and off sides. This phase of instruction of the young horse should be carried out very thoroughly. Absolute calmness must be obtained even in the midst of noise and movement—such as might be expected in war, when it might be most valuable to have a perfectly quiet horse to mount. However, it is well not to require too much at the beginning.

82. MOUNTED WORK.

Out-of-doors and in the Riding Hall. As soon as the horse accepts the rider, his conditioning can be carried forward more rapidly. This work of conditioning becomes of paramount importance and must progress through the periods of breaking and training without interruption. It should take place out-of-doors whenever possible.

The remount working in the fresh air and on straight lines naturally acquires its full strength most rapidly. Nevertheless, the first sessions should take place in the riding hall to enable the instructor to exercise closer supervision, to give him a better opportunity to study his men and horses, and to avoid the greater possibility of accidents outside. The riding hall is used to give the remounts the first lessons in the aids.

At the start, some quiet old horses mixed in with the remounts may have a very good influence on the entire section.

83. ELEMENTARY LESSONS IN THE AIDS.

Before taking the young horse outside, he must understand and be responsive to the signals for four different actions. All possible movements and all gaits are derived from these four elementary actions, and it is necessary that they be taught as soon as practicable, and separately, before they are associated in producing combined effects.

These actions are: To move forward at the pressure of the legs; to slow up or stop at the tension of the reins; to turn to the right and left with opening reins.

These elements being thoroughly understood by the young horse, the ordinary exercises of equitation judiciously adapted to his means, to the maturity of his development, and to his physical condition, will accomplish his education.

The accomplishment of these latter (the ordinary exercises of equitation) will be the role of training properly speaking. The horses are bitted with large single snaffles. Special attention must be given to the use of proper bits, correctly adjusted.

84. TO MOVE FORWARD.

The basis of all training is freedom in the forward movement.

It is necessary then, from the beginning to teach the horse to respond to the action of the legs. This lesson is the first one given and should be repeated frequently. For the first sessions the following rules are appropriate:

a. Do not let the legs remain glued to the side of the horse; act by repeated impulses.

b. Touch the horse near the girths; not too far in rear.

c. Begin by giving this lesson in passing from the walk to the trot, then in extending the trot, finally in passing from the halt to the trot.

d. Reenforce the action of the legs, if necessary, by following these actions immediately by the use of the voice, or even by light taps of the whip. This last suggestion is more particularly applicable when the lesson is given in the riding hall.

Outside on the road, and especially when marching behind an old leader, the remounts have a natural tendency to move forward in order to follow. That is another reason in favor of working out-of-doors at an early date.

In the forward movement the reins should always be stretched. Otherwise, instead of being enclosed by the aids, the horse is uncertain of his direction; he wobbles from side to side, and the rider cannot direct him.

It is easy to maintain the reins stretched with energetic horses which have natural impulsion; it suffices for the rider to maintain steady hands, and without discouraging the spirit of the horse, to gradually moderate his excess of ardor. It is more difficult to teach lazy, cold-blooded, or grass-raised horses to go up to the hand. These, as a rule, only stretch their reins when tired, in order to support their heads. With such horses, from the very beginning, the rider must seek the mouth. Later on, strengthened by regular work, the remount having learned the habit of taking contact with the bit, when urged by the legs, will stretch the reins of his own accord. When the horse obeys the action of the legs, work on straight lines at the walk, trot, and gallop will lead him gradually to take the desired support on the hand, hence the rider should be careful not to discourage him by undue severity.

A set hand with fingers closed would only confuse the horse, and, being painful to the mouth, would have the effect of sending him back from the bit.

Thus, in the first lessons in the forward movement, the hand should not oppose the extension of the neck; the fingers, on the contrary, should be ready to yield in order that the neck may stretch out and that nothing may interfere with the willingness of the horse in his movements forward. The legs are active, the hands passive.

With some particularly cold-blooded horses, who obstinately remain deaf to the call of the legs, it may be well to use the spurs at an early date. But even in this case the spurs should have dull rowels, or preferably be without rowels.

With well-bred horses it is the rule not to use spurs during the first lessons.

The use of the spurs in most cases does not necessitate a special lesson; the horse nearly always responds to them by a bound forward.

With mares or balky horses, that kick at the leg and refuse to advance, it is generally sufficient to put them on the longe and make a vigorous use of the driving whip coincident with the rider's attack with the legs.

85. HALTING.—To halt, the rider progressively closes his fingers, and slightly moves the upper part of his body to the rear. He opens his fingers as soon as the horse responds.

He regulates the severity of his hand to the degree of sensibility of the horse's mouth.

In halting, the horse should remain straight and up on the bit.

Nervous horses, those that engage their haunches too greatly and those with a tendency to back, should not be halted frequently. On the other hand, those that, from their conformation, have too much weight on the forehand should be frequently halted. Training is nothing more than a search for balance, and the halt is an excellent gymnastic for those horses that, having a high and powerful behind, are difficult to slow down.

86. THE TURN.

Horses are accustomed to being led in a snaffle bridle; this familiar opening effect of the rein, which is always accepted by them, serves as the starting point in their education.

To turn to the right, the rider opens the right rein gently, carrying the hand forward and to the right. The other hand, which does not act, must be carried forward and low so as not to contradict the active rein.

It is very important that all sensations which the remount is made to experience should come to him distinctly and clearly.

The opening effect of the rein should be produced laterally, and with the least possible action from front to rear. It provokes, nevertheless, a slight slowing up, which should be combatted by increased action of the legs.

When the remount is well confirmed in the forward movement, in order to confirm and fortify the action of the leg, advantage is taken of the fact that the action of either opening rein tends to have the same result as an active leg on the same side. The remount is thus taught the action which results from the opening of one rein and the closing of one leg, and he is habituated in carrying his haunches away from the acting leg.

When the horse readily obeys the action of the opening rein, at the walk and trot, he is taught the action of the bearing rein. In view of the gymnastics to which he later will be submitted when he is guided by one hand, he is taught to turn to the right, for example, by action of the left rein. To obtain this result, it is well to take advantage of the corners of the hall, or exercise the horse in turns by the flank, abouts, or other circular movements with the opening effect of the rein; and as soon as the horse commences to obey, to substitute the action of the bearing rein by immediately carrying the left hand (for a turn to the right) forward and to the right. The opening rein serves as a sort of interpreter for the bearing rein. As soon as the latter comes into play the action of the opening rein should be discontinued, and the right hand lowered to allow the bearing (left) rein to produce its full effect. After repeatedly alternating the effects of the two reins closer and closer together, the action of the opening rein is diminished, and then gradually eliminated, as the horse comes to understand more clearly what is being asked of him.

87. FIRST LESSONS AT THE GALLOP.

The importance of the gallop requires that the rider be familiar with all details of this gait, for the manner in which the horse takes it, keeps it, modifies it, or leaves, it has great influence on its value. Furthermore, this one more or less complicated movement brings out all the rider's skill and requires implicit obedience on the part of the horse. In obtaining it the rider may apply all the principles set forth in training and he may demonstrate in detail the role and value of the aids. To analyze completely the gallop departs is to sum up all training. The study of this one movement will show how all others are obtained.

The horse at liberty takes the gallop in different ways according to the circumstances which provoke it; that is, by a sort of loss of balance in throwing his weight forward, or by engaging his hocks under the mass in balancing himself.

Mounted, he acts in the same way when he takes the gallop voluntarily and not in response to action of the aids, as by a cluck of the tongue, crack of the whip, fear, etc.

When the rider desires to obtain the gallop by use of the aids he must consider the effects which the hand and legs may produce. The partially trained horse can understand only a part of the effects of the aids. In order that the horse may understand that the sensation he receives calls for the gallop, he must first be placed in such a position that all confusion and hesitation are removed and only one movement is left to be executed—the one demanded. Position should always precede action, regardless of whether the horse takes the gallop through loss of balance or while in perfect balance.

The very mechanism of the gallop indicates the position which the horse should be made to take. The gallop is characterized by one lateral pair of legs being more advanced than the other; thus in the right gallop the two right legs are more advanced than the two left legs, and vice versa.

Young horses may be ignorant or imperfectly disciplined to the aids, but they must nevertheless be galloped for their development and to advance their conditioning. The horse should be started on some circular movement (circle, passing through a corner, movement by the flank) where the interior lateral biped, having less ground to cover than the exterior biped, may easily be advanced more than the latter. By pushing more or less vigorously with both legs, when the horse is thus placed, the gallop will be taken naturally, especially if the rider carries his body forward and slightly to the right for the right gallop. The horse is, so to speak, surprised and thrown forward by the leg action. He more or less *falls* into the right gallop.

Since the reins do not have to act on the horse's balance, this method has the advantage of leaving his head free, and of putting him in the gallop before he has realized it, and at the same time, leaving him calm and preserving his impulsion.

The rider has profited by the favorable position taken by the horse himself to obtain the gallop. By this method the horse is induced to gallop and is familiarized with the gait under the rider's weight; by practice the departs become easier and the horse goes into the gallop willingly.

This method represents the first step in the lesson of the gallop. It is sufficient to give the remount the necessary work at the gallop. In fact, it is the only one which may be employed at this time since he is ignorant of even the elementary actions of the aids.

88. PRELIMINARY CONDITIONING.

The proper conditioning of remounts is a most important part of their training. In order to condition a group of remounts, the instructor must plan ahead and set himself a definite goal consistent with the time allowed. Then, by constant supervision and strict adherence to a prearranged schedule, steady progress can be made.

The first and perhaps most important factor in conditioning is proper stable management. Regular feeding and watering, thorough grooming and careful attention to minor details will result in healthy animals. Careless tying on the line so that horses may kick each other, inattention to the care of the feet, and similar common faults result in loss of training time and a consequent failure of the remount to be properly conditioned in the required time.

The second factor in conditioning is the actual work given the remount. For the first few days outside work should be done in a group at the walk. If the group goes quietly the instructor should divide the groups into pairs, with at least one of the horses in each pair a quiet horse. Then as they become accustomed to the strange noises and sights outside, individual outside work may be started.

In group work or marching in column, the instructor should pay particular attention to the rate of the gaits. Too slow a gait will result in horses becoming upset due to their continued restraint, and too fast a walk will cause the slow horses to jig. When a particular horse cannot conform to the regulation gaits, he should be taken with the group but should be worked individually or with another horse until he can conform.

Condition cannot be attained by riding hall work alone. The work should be varied so that the horse does not become routined.

The amount of work should be gradually increased as the condition of the group improves. Long gallops or road marches should not be attempted until the general condition of the group indicates that the individual horses are up to the work. A schedule, which does not progressively build up the muscles and wind, is bad. *Consistent progressive work is the keynote to conditioning.*

The choice of ground is an important part of the outside work. Remounts not yet conditioned are easily blemished and injured by work on rough or hard ground. At first all work should be done on the softest footing available. As their condition improves, short trots on hard ground should be given in preparation for road marching. Later, when the remount has acquired balance and strength, taking him over varied ground is good practice and necessary to his training.

When the remounts are being returned to the stables, it is advisable to select a different route each day, otherwise they may become fidgety and nervous in their desire to return. It is very important that the remounts be returned to stables dry and cool, and in condition for immediate grooming.

89. VALUE OF THE VARIOUS GAITS.

The *walk* plays an important part in conditioning because it can be sustained for a long time without fatigue. At its full extent it supples all points, strengthens and hardens the tendons, and engenders calmness and power. By accompanying the play of the neck, which is very pronounced at the extended walk, with the hands, the horse gains confidence and develops the habit of keeping contact with the bit. For these reasons, this gait should play a very large part in breaking.

The trot is useful in the beginning, at first for "taking the edge off," then for putting the remount in the forward movement while teaching him to accept the bit, which is necessary for him to march straight; thus enclosed by the legs, which push him forward, and the hand which supports him, the horse acquires the habit of fixing his neck in the direction of movement, which facilitates his later training.

From the physical point of view, the trot stimulates the circulation at the same time that it develops the muscular system.

The periods at the trot, in the beginning, should be frequent and short. The length is increased as condition improves.

The gallop is the gymnastic par excellence for the young horse; it puts him at once both on the haunches and on the hand, and it develops his respiratory facilities to the highest degree. It is a gait which the horse should be able to sustain for a long time without fatigue; therefore training for it must be started early, though because of the mechanism and power of this gait, it should not be used outside except on good ground. Lacking favorable ground, it will be preferable to gallop only in the riding hall, until the horse is more developed.

Outside there is no question of proper leads; the rider proceeds as in the riding hall, he regulates the length of the gallop by the progression of his work, and above all by the physical condition of the horse. These gallops, at first very short—400 to 500 yards—are increased progressively to reach 1,500 to 2,000 yards at the end of the period.

In the beginning, the bringing on of a struggle which might be harmful to the remount's mouth, interfere with his gaits or kill his natural impulsion in order to regulate the gait, should be avoided. The objects in this part of the work above all are the physical development and the establishment of confidence. The rider plays a passive part, so to speak, and makes every concession which will not have bad results on the health or character of his horse.

90. DEFENSES OF THE YOUNG HORSE.

Here a distinction must be made between meanness and playfulness. While the first should be suppressed from the beginning, it would be wrong to punish the impetuousness of the remount. When the rider feels that the remount is about to cut up, he should close his thighs and lower legs, steady his hands, close his fingers, and wait. Likewise, when the remount escapes the hand and bolts to the front, or jumps to one side, no attempt should be made to control him by a regular use of the aids, as he barely understands them when he is calm and moving at slow gaits. Therefore, the rider should steady his hands, use his voice until the remount is calm, stop him, face him in the proper direction, and move him forward. This procedure, even with old horses, always gives the most satisfactory and prompt results.

Under any circumstances, the instructor cautions patience and gentleness.

CHAPTER XIV
ELEMENTARY TRAINING

91. General	173
92. Principal Factors in Training	175
93. Influence of Character on Training	176
94. Limit of Training	176
95. Use of the Curb Bit and Snaffle	177
96. Basis of an Equestrian Language	178
97. Resistances	180
98. Gymnastic Work for the Young Horse	181
99. Principles of Movement	181
100. Extending and Collecting the Gaits	185
101. The Halt	186
102. To Change Direction	187
103. To Obtain the Engagement and the Mobility of the Hind Quarters	189
104. To back	190
105. To obtain Free Play of the Shoulders	191
106. Summary	192
107. Conditioning	192
108. Use of Arms	193
109. Preparation for Field Service	193

91. GENERAL.

The trained horse is one that understands the rider's intentions from his slightest indications and immediately responds to them with exactness, lightness and energy; in other words, he is sensitive to the leg and light on the hand.

Training is distinguished from breaking in that, during the breaking period, acclimatization and physical development are of the utmost importance, and the rider must at this time make many concessions to the young animal. However, during the training period the horse must submit to the demands of the rider and must give him complete obedience.

Training should not be started until the horse, strengthened by fresh air and rational conditioning, is free in the forward movement, has confidence in his rider, and is in shape to learn the language of the aids and to subject himself to their requirements.

To obtain good results, training must be based on a sound doctrine; it should follow a method, and should conform absolutely to the rules of both.

A *doctrine* is a group of principles whose validity has been established by reason and justified by experience.

The fundamental principles of training concern themselves with the development of a calm, forward moving, straight and manageable horse.

The *method* of training prescribes the various means of execution and regulates the order of their use. In application it varies according to the particular end in view and to the circumstances of time and place. It is influenced by the temperament of the instructor and the skill of the rider.

The method of training herein set forth has as its objectives:

a. The development and exploitation of the horse's physical strength and moral qualities.

b. His submission to the aids, obtained by a rational and progressive education and without the use of force.

c. The search for balance.

The progressions of the training as suggested herein does not include rules or manner of execution. Wherever movements or figures are mentioned, their execution is not entered into in detail.

The training value of such figures depends entirely on the manner in which they are executed; it is the position imposed on the body of the horse by the rider's aids which makes them of value. A rider may execute all of the figures prescribed in the regulations in the most approved sequence and never obtain the result sought for. Another, working on these same figures but with a clearly defined objective, and using his aids towards the attainment of this objective, will train his horse very quickly.

92. PRINCIPAL FACTORS IN TRAINING.

a. The Instructor.

The ability of the *instructor,* as well as that of the *rider,* plays a very important part in training. The instructor should possess a profound knowledge of the horse and of the theoretical and practical science of equitation. He should have a thorough understanding of the method of training. In addition, if his explanation is not sufficient, he must be able to demonstrate, in a skillful manner, any point in question.

b. The Rider.

The personal ability of the rider is, however, the principal factor in the training of the horse. However excellent may be the method of training, the quality of the horse, or the capability of the instructor, the horse will never be wholly obedient if the rider does not thoroughly understand his part. The primary requisite, therefore, for the training of a horse is a skillful rider.

c. The Work.

Any method of training not based on the conditioning of the horse is not a good method. In training, as in breaking, outside work will aways remain the essential part. The instructor must maintain the proper proportion between outside and riding hall work. Outside work is necessary in order that the conditioning of the young horse can keep pace with his training. Aside from conditioning alone, outside work is necessary to assist in training the remount in all the work that a well-rounded military horse must perform. Among other things, a military horse should go calmly and freely at all gaits cross country, either alone or in a group, he should jump obstacles of all types without excitement, he should be able to bear arms and remain quiet during all types of firing and confusion about him, and he should be able to carry his rider at steady gaits on road marches. If the horse is to be trained in these essentials, the instructor must so arrange his schedule of training that they all receive their proper proportion of time throughout the training period. By a proper coordination of training, discipline, obedience, and lightness, inculcated in the remount as a result of the gymnastic exercises of indoor work, can be utilized to perfect the young horse in his outdoor training and assist in adding balance and control, which are indispensable.

Under normal conditions the work in the riding hall should not exceed a third of the whole; and the sessions therein should not exceed an hour each, during which time frequent rests should be allowed.

d. Time.

One should not forget that whatever may be the rider's skill, the physical and mental development of a remount varies with each individual. No scheme of training can hasten the natural growth and development of the remount or substitute itself for the work of time. Patience and gradual increase of effort are most effective in training; to go slowly is, in this case, to go quickly.

93. INFLUENCE OF CHARACTER ON TRAINING.—The horse's mental make-up has much to do with his education. It must be studied in order to take advantage of it.

The chief mental characteristic of the horse is his memory. This faculty aids training if proper use is made of it. On the other hand, it makes errors dangerous; it is difficult to retrain a horse—that is, to make him forget bad lessons.

Natural aptitude varies in different animals as do the qualities of intelligence. Some understand at once what is asked of them, others are very slow to learn.

The horse is usually docile and tractable; he is sensitive to good treatment, to the voice and caresses. Blows will not subdue him; they only serve to make him irritable and suspicious.

He is capable of attention and reflection, since often he will execute in the morning movements which were difficult for him the evening before. He is imitative, and it is on this trait that the use of leaders is based.

He is patient, but his patience is not unlimited. One of the difficulties in training is to determine the limit of requirement that the horse will endure, and profit thereby during each period.

Training disciplines the horse, but cannot completely transform his character. A mean or tricky horse, however well trained, is always to be mistrusted. The rider may hasten the submission of the horse by studying his mental faculties and taking advantage of them.

94. LIMIT OF TRAINING.

The instructor studies each individual horse, decides the means of training which best suit his case and regulates the work accordingly.

It is not possible to bring all horses to the same degree of perfection; but one should always endeavor to develop the full capabilities of each horse, without however, trying to obtain by force that which the animal lacks the strength or ability to give.

There are some methods of training that aim to subdue, finally and completely, the horse's mental and physical forces, and thus obtain absolute domination over him. These methods are based on the complete collection on the spur, but they are not in the domain of secondary equitation. However, the work required of the military horse does make a constant demand on the horse's instinct and initiative.

95. USE OF THE CURB BIT AND SNAFFLE (ALSO KNOW AS THE BIT AND BRIDOON).
a. General.
When the remounts give themselves over freely in bold and extended gaits, and when they accept the support of the hand on a snaffle without hesitation, the double bridle may be put on without fear, remembering always to demand nothing but work on straight lines until the bars of the mouth are accustomed to the bit. Thus the danger of letting the remounts acquire the bad habit of overloading the shoulders and boring on the hand is avoided. The transition from the single snaffle to the double bridle may be made by giving the horse several periods of work on a double snaffle thus accustoming the horse to carry more bitting in his mouth.

In training there is no regulation way of holding the reins. The rider must determine, according to his objective and the resistance he meets, the method of holding which permits him to utilize most effectively the appropriate actions of curb and snaffle.

b. Adjustment of the curb bit and snaffle.
When first used on a remount, the curb bit and snaffle should be most carefully adjusted in such a manner to protect the unaccustomed mouth of the young horse from injury. With this in view, both bits should be placed high in the mouth and the curb chain made quite loose. The rider, moreover, should use the snaffle bit almost exclusively. The curb bit should be brought into play very lightly, gently, and gradually. After a few days the bits and the curb chain may be adjusted more correctly.

c. Accustoming the young horse to the effect of the curb bit and snaffle.
During the first few day's work with the curb bit and snaffle, it is better to go chiefly out of doors. Any work indoors should be mild in character, nothing of an exacting nature being demanded of the horse until he has fully accepted and goes well into the bit and bridoon. There is now great danger that the young horse may "get behind the bit." This must be most carefully guarded against by the officer in charge and by each rider. In all exercises and work

the reins should be held lightly stretched and quite long, a lighter pressure on the curb than on the snaffle. The horse should be encouraged, by tactful use of the legs, to extend his head and neck to the front and to maintain a steady pressure against the bit. If he is not inclined to do so, the rider should not shorten the reins or draw back the hands; he should hold the hands still and use the legs in a manner to preserve brisk forward movement and thus bring about extension of the head and neck and steady pressure into the bridle. During periods of rest, whether standing still or while moving, the reins should not be held taut; they should be allowed to swing loose, thus according to the horse full liberty to relax, to extend himself and to rest. Horses grow restive if kept too long in a constrained position. They should never be so held during movements of relaxation or periods of rest.

96. BASIS OF AN EQUESTRIAN LANGUAGE.

In order that man may transmit his wishes to the horse, make use of the animal's intelligence, and thus dominate him, it is necessary to establish between them a sort of conventional language which the rider may easily learn and utilize, and which the horse may as easily understand and accept.

This language is based on the law of association of sensations;

"When impressions have been produced simultaneously or have immediately succeeded each other, it is thereafter sufficient that one of the impressions be presented to the senses in order for the others also to be called to mind."

For example, if a horse moves forward at the cluck of the tongue, it is because one day he saw a whip, felt the lash, and at the same time heard the cluck. Whenever the last sensation, affecting only the hearing, is alone presented, those of sight and touch will come immediately to his mind ,and he will move forward as he did under the effect of the lash.

Likewise, the horse which has learned to range his haunches under the effects of the riding whip, later will range them under the pressure of one leg, because the two sensations will have been associated in the beginning.

The movements that the horse executes naturally under the influence of the aids are very few. The most docile horse cannot obey his rider's commands if he does not understand them. It is by building upon the principle of the association of sensations or ideas, cited above, that one may construct the language which makes for this indispensable understanding. Sight, hearing, touch, and even taste successively come into play, and each has its part in this education.

The elements of instruction are given on the longe. The touch, then the sight alone of the whip, produces the forward movement. With this should be associated the cluck of the tongue, for which, later on, is substituted the action of the legs. In the same way the traction of the longe paves the way for the opening rein, which in turn will serve to introduce the bearing rein.

From the opening and bearing actions of the reins, the horse will come at length to understand the actions of the reins of opposition. The actions of the legs will follow, and finally, he will learn to respond to various combinations of actions that cannot be exactly defined but which depend on the tact and finesse of the rider.

It is apparent that there exists from the beginning the need for the greatest clearness in the impressions transmitted to the horse. It is on the distinctness of these first indications that the clarity of the entire language will depend, and in consequence the rate of progress of the animal's education. But it is not sufficient that the horse grasp the meaning of the demands made upon him; he must also have the pliancy of will to carry out the often distasteful requirements that his rider imposes.

Here again it is the law of the associations of sensations which offers the means of assuring the horse's obedience. When a well executed movement is rewarded at once and an act of disobedience immediately punished, the horse soon learns to submit to the rider's demands.

The continuous repetition of this procedure will gradually result in a more and more prompt obedience and finally to an absolute and instinctive submission.

To attain this last result, training requires not only gentleness in order not to irritate nervous horses, but also much firmness, for the horse must feel that his master is possessed of infinite power; his submission depends upon it. During the course of training, the time always comes when a difficulty appears, when resistance is offered. The tactful rider determines whether it is due to physical inability or unwillingness. In the first case one must be particularly patient and moderate in one's requirements. In the second case, on the contrary, one must take up the struggle resolutely and emerge victorious lest the horse, by association of sensations, becomes conscious of his strength, and, in consequence, obstinate.

Moreover the trainer must not be deceived into believing that the horse is disciplined because he appears to be resigned. *The trainer should be keen enough to foresee the warning signs of impatience and revolt, and to modify or cease his requirements in time.*

A horse may be easily prevented from becoming impatient, and yet made to repeat the same series of exercises daily, provided the work is varied; and provided he is given brief periods of rest after well-executed movements. These rest periods lessen the monotony and afford the horse needed opportunity to relax.

In order that the horse's education may be complete, his obedience must be more than prompt and absolute; it must be automatic. At this point, the indication of a single one of the aids heretofore, strictly combined with others, suffices to start the mechanism of association and produce the execution of the movement demanded. In the beginning, in order to obtain even a labored gallop depart, it was necessary to use both hands to displace the forehand and both legs to give the position and inspire the impulsion; later, closing the fingers on the rein or the mere contact of the boot suffices to obtain this same movement because this action has awakened the remembrance of all the other previous sensations.

It is by repetition that associations penetrate the memory; the operation is necessarily one of long duration. But by substituting for the repetition, or rather by adding to it, the intensity of a transmitted sensation, the progress is hastened. Strong impressions, even though seldom repeated, engrave associations in the mind more quickly than weak ones which may weary or enervate the horse.

In accordance with these principles the curb bit and spur when properly used expedite training.

If the horse, through distraction, laziness, or bad intent, attempts to avoid doing what is expected of him, the energetic action of the fingers on the reins or a simple application of the spurs will quickly remind him of the established conventions.

Fixing the associations by the intensity of one of the associated impressions is one of the fundamentals of training.

97. RESISTANCES.

The fatigue caused both by the work of training, and the constraint imposed upon young horses during their education, may occasion certain resistances. These are most apt to occur in horses having physical defects, blemishes and weaknesses, or on account of nervousness arising from awkward or misunderstood demands.

Whether the cause of these resistances be mental or physical, the joints and muscles may be strengthened and suppled by appropriate gymnastics applied to the various parts of the horse's body. The principal points of resistance are the haunches, the spinal column, the shoulders, and the mouth. Certain movements lend themselves more particularly to the suppling of certain parts; the instructor regulates their use according to the end in view.

98. GYMNASTIC WORK FOR THE YOUNG HORSE.

Gymnastic work tends to develop the remount's strength and suppleness, and it also serves to conquer any resistance he may offer.

Gymnastic work is practiced at all three gaits; at the walk, because the rider is steadier and has greater control over the horse, while the latter is more attentive and better able to understand the movements demanded; at the trot, because from the conformation of the horse it is at this gait that the joints play most easily in lateral movements; and at the gallop, because it is the combat gait, and consequently the ultimate objective of training.

99. PRINCIPLES OF MOVEMENT.

a. Impulsion.

In equitation the locomotive energy of the horse is called *forward* movement or *impulsion*.

The *forward movement* is the first degree of impulsion. This quality exists in the horse when he responds to the first pressure of the legs by extending his action without sensibly increasing the height of his movements.

Impulsion is forward movement subjected to the exact discipline of the aids and exploited in view of the object to be attained. It is the basis of training. Its origin is in the hind quarters which propel, or are ready to propel, the mass forward.

Impulsion may be either natural or acquired; it is the natural instinct in warm-blooded, generous horses; in the cold-blooded or lazy horse it is obtained as a result of training in response to the aids, and dies out as soon as the aids which provoked it are relaxed.

The rider is fully master of his horse only when he is master of the horse's entire impulsive forces. Certain horses are unwilling to recognize this mastership and often oppose it by a most complete inertia; others use their whole muscular power to struggle against the rider, to resist him, or to escape from him completely, but most horses lend themselves generously to the rider's wishes, and seem to put all their strength at his service. It is this moral and physical submission to the aids in the forward movement which should be sought above all else in training.

Speed is not a criterion of impulsion. Impulsion is shown much more by the manner in which the horse lends himself to the rider than by the rapidity of the gaits. One horse, marching at the trot or gallop, even extended, may lack impulsion, while another may show a great deal of it at the walk.

This freedom in the forward movement should be carefully guarded, not only during training, but during the whole military life of the horse.

Proper use of the horse's strength leads to proper division of weight, and consequently in turn to balance, mobility and tractability.

b. Balance.

The horse's muscular force and his weight are the two elements which unite to produce movement.

Muscular force is essentially productive of energy. It causes the displacement from inertia and its use, therefore, determines whether the distribution of the weight is proper or not.

The object of training is to govern this force at the several gaits, at all degrees of speed, and in the changes of direction in such manner as to oblige the horse to execute the demands of the rider.

Theoretically, movement is determined by the various positions of the center of gravity with respect to the base of support. In the state of rest the center of gravity is sustained by that base. Movement is but the disturbance of that equilibrium, the members intervening to steady the mass and prevent a fall. Thus the four basic movements—forward, backward, to the right and to the left—always take place because the center of gravity draws the mass to one of these four directions.

In practice, a balanced horse is one that is light in his gaits and agile in his changes of direction.

The horse at liberty balances himself naturally. His movements are more or less easy, and the mobility that he exhibits shows that he is master of his strength and can make judicious use of it. With rare exceptions, as soon as the horse is mounted, this natural equilibrium is disturbed by the rider's weight, which displaces the center of gravity and usually two-thirds of this weight falls on the forehand. In addition, the voluntary or involuntary actions of the aids provoke numerous contractions so that a part of the horse's muscular power is employed in resisting the rider.

The less the horse resists his rider, the better he can balance himself and the more manageable he becomes.

The capability of a horse to make proper use of his strength depends upon his conformation. A well-made horse balances himself more easily under the best possible mechanical conditions. But irrespective of the conformation of the horse, the rider should endeavor without delay to make him regain a natural balance, or at least a balance which approaches it. Therefore, at the beginning of his training the horse must be allowed great liberty, for if his movements are restrained he will be unable to recover this balance. The more steady, exact, and conciliating the rider is in his actions, the more confidence the horse will have and the quicker he will regain his equilibrium.

As training progresses the number of resistances will diminish; the horse in yielding to the aids will make better use of his strength and a better distribution of his weight; and the rider will find less difficulty in giving him the position leading to the desired movement.

c. Locomotion.
The laws of locomotion are concerned with the order in which the horse places his feet in the different movements and in the several gaits.

In superior equitation that application of these observations may lead to good results; but in secondary equitation a wider view must be taken lest the difficulties be aggravated by trying to attempt objectives not wholly practical under the conditions.

The rider need only concern himself with giving the horse the position which should precede each movement, leaving to the horse the care and time of placing his feet accordingly.

d. Role and position of the head in movement.
When the horse uses all of his natural means to aid himself in the execution of his movements, he employs his head and neck to balance or modify the employment of his forces. If he wishes to move forward he straightens his head to the front and stretches out his neck so as to draw the center of gravity in the direction of the movement. Or, if he wishes to stop or back, he brings in his head, shortens his neck, and thus impels the movement of the mass to the rear. In movements to the side, oblique or circular, it is again the displacement of the head and neck to the right or left which facilitates, regulates and maintains the movement.

The rider who wishes to be master of his horse should place the horse's head in such a position, by means of the bit, that he may regulate its displacements, and, in turn, those of the neck. In this way the neck bends, shortens or stretches out, according to the impressions which the mouth receives from the rider's hand.

In order that the impression of the hand may be clearly transmitted to the horse's mouth and at the same time that the breathing may not be hindered, the horse's head should slope from the muzzle upward and rearward and it should be made to take this position at the ordinary gaits and in the simple, regular movements.

The more shortened the gait, the more the head should approach the vertical line; conversely, the more it is desired to increase the speed, the more the head should approach the horizontal. In both cases the position the head assumes is considered normal since it favors the slowing or the increasing of the gaits respectively.

The head may assume an exaggerated position (too close or too far from the vertical) due to defective conformation of the forehand, badly adjusted bitting, excess of sensibility in the chin groove or bars—and as is most frequently the case with horses that "stargaze"—because of defective conformation in some part of the hindquarters.

The rider combats the faults of position not only by a judicious use of the aids, but also by the use of a milder or more severe bit; by raising or lowering the bit in the horse's mouth, and by loosening or tightening the curb chain.

Thus with a "stargazer" the head should be brought in by increasing the lever arm; that is, by using a long branch bit placed low in the mouth. The horse that carries his head too low, or that is too much gathered, should be fitted with a short branch bit adjusted as high in the mouth as possible.

The position of the head in the natural state is determined by the posture of the neck. When the horse is under control, the bit, with its action on the bars of the mouth, makes the head take a position to which the neck is forced to yield. The neck will therefore contract, stretch out, or bend to the right or left, according to the action of the hands.

e. Role and position of the neck.

The neck is the indispensable lever for facilitating movement. Its position and its attachment to the head should be such that, while it remains supple and accepts the lateral and retrograde displacements without resistance, it should always preserve its natural stability, and even a certain degree of firmness at the base and up to its middle portion. Its direction should be identical to that taken naturally by the horse, unmounted, when he is placed in the position of attention.

If the head and neck are raised too much, the play of the shoulders will be freer, but at the same time the loin and the whole hindquarters will be cramped, the haunches and hocks hindered in their action, and the displacements of the hind quarters restrained, unequal, and jerky. Consequently, the gait will lose speed and regularity.

If the neck is too low, the overcharged shoulders will render the horse heavy and difficult to guide.

Thus the neck should be neither too high nor too low. It should shorten or stretch out as the head approaches or leaves the vertical. When coming in, the head should cause the neck to bend at the poll without breaking its line; in stretching out, it should extend the neck without raising it.

With the head and neck properly placed, the reins will retain their full force, and the members of the forehand like those of the hind quarters will coordinate their actions in the movements, whether collected or extended.

In determining the position to give the neck, the rider should always take into consideration the manner in which it is naturally joined on (the "set on"). Some riders make the mistake of demanding great elevation of neck from horses that naturally carry low heads. In attempting to raise the neck too high for the natural conformation of the horse, the hand often stops the impulsive forces. When a position contrary to the horse's conformation is demanded, the hocks and loin are cramped, and there is a resultant loss of freedom in the gaits.

100. EXTENDING AND COLLECTING THE GAITS.

This work includes:

Being at the walk: Slow walk, walk, extended walk; from the slow walk change to the extended walk, and vice versa.

Being at the trot: Slow trot, trot, and extended trot; changing from slow trot to extended trot, and vice versa.

Being halted: Walk and halt; extended walk and halt. Being at the walk: Taking up the extended trot; halt from the trot or extended trot.

Finally it includes the gallop departs from the trot, the walk, the halt and from backing; passing from the gallop to the trot, to the walk, halting from the gallop, extending and collecting the gallop.

During *breaking* the horse learns to obey the hand by slowing the gait, and to obey the legs by an increase of gait. This is again taken up and insisted upon until the slowing of the gait is obtained without the least movement of the head—which would indicate a struggle against the hand—and until the extension is very willing and immediate. Each time the horse fights the hand while slowing down from faster gaits, he must be put back to slowing the walk.

Obedience once obtained, attention should be paid to the *manner* in which the movements are executed; the haunches should be drawn under the mass in slowing, and the action of the hocks and loin in the extension should be vigorous. This is achieved by alternate extensions and collections, though nearer and nearer together.

After these results have been attained, one should demand the most marked extensions and collections; for example, immediate halts from fast gaits.

In this work of balancing the horse between the hands and legs, *it is essential that these two aids never act simultaneously.* In slowing down, the horse should let the gait die out while moving close to the ground.

At the indication of the legs, he should push himself vigorously forward. If his motions are high, it is because the hand has not yielded in time to let the impulsion pass.

If the horse moves sideways in slowing, he is straightened by opposing the shoulder to the haunch. These supplings are interspersed with periods of work on the bit at free gaits. With lazy horses one should insist especially on the promptest obedience in the extensions, and with high spirited horses, in the collections.

When once this work has been well-executed on the straight line, it is repeated on the circle. This requires a relatively greater engagement of the interior haunch. The diameter of the circle is reduced in proportion to the progress of the horse. But the horse himself must not be allowed to modify this diameter in accordance with the gait, that is, to make it smaller in slowing and larger in extending.

Inside the riding hall the increase and decrease of the gallop should be confined to speeds appropriate to the enclosure (not too extended). Outside, on straight lines, the speed should be varied through every form of gallop from the most collected canter to the run, and the reverse. This work well done is proof of the thorough execution of all preceding work.

If the horse bores or fights the hand, he must be returned to the riding hall for work in slowing the walk and canter; the object is thus accomplished more surely and with less risk to his legs than if one were to insist on submission in the work at fast gaits.

101. THE HALT.

Considered from the point of view of training, the object of the halt is not to stop the horse in some position or other, but rather to teach him to balance himself by the engagement of his haunches.

The halt is effected by the action of the fingers closing on the adjusted reins. If the hocks remain out in rear or are thrown to one side to avoid an engagement, which is always difficult at first, the legs intervene to push the haunches gently under the mass. The hand remains passive.

Exercises in halting serve to gather the horse's forces to place his head and haunches and to make him lighter.

The halt must be practiced gradually and very gently at first. One should be very careful with horses whose necks are "upside down," who are long coupled, sway backed or nervous. Horses of this kind are generally weak in the loin. They halt with difficulty, and the remedy becomes worse than the evil. Halting exercises must be avoided with horses that are lacking in impulsion.

Summing up, the halting exercises are suitable only for horses having good loins and sufficient vigor in the haunches, and hocks, to stop easily on the hindquarters.

This does not apply to the *half-halt*—defined in Paragraph 26 which irritates the horse less than the halt. One may therefore employ it without fear on all horses except, again, those that hold back. The result sought is to lighten the forehand, and consequently to lower the croup, without producing a slowing of the gait.

102. TO CHANGE DIRECTION.

a. *The turn.*

The turn comprises two distinct, but closely united operations: the change of direction and the march.

It is assumed that the reins are adjusted at the moment the turn is to be executed, as they should be at the commencement of any movement whatsoever. This being the case, the rider has several means of displacing the forehand to determine a change of direction. The simplest means, by which the beginner can cause his horse to make required changes of direction, consists in employing the opening rein by carrying the right hand to the right, if a turn in that direction is desired. This draws the head to the right, and the shoulders and the rest of the body follow, provided the horse does not resist. The change of direction is thus executed as well as can be expected. One of the disadvantages of this method is that it may displace the haunches to the outside which cannot have other than unfavorable results.

In addition, the turn by means of the opening rein has the inherent defect of this rein, which is to make it possible for the horse to yield his head and neck and not follow them with the shoulders. The rein aids thus are valueless, and the rider no longer has control of the horse.

The turn to the right may also be accomplished by the employment of a left bearing rein. This rein, when used alone, has the same inherent defects as the opening rein, in that it is possible for the horse to yield its head and neck and not follow this direction with its shoulders.

Another method employed for a turn to the right is the use of the left indirect rein of opposition in front of the withers. It produces the change of direction and is of great assistance when riding with the reins in one hand. In acting alone, however, it weights the right shoulder more or less strongly, opposes this shoulder to the right haunch, thus slowing down the horse and necessitating a more active use of the legs than is required when the turn is made in a more rational manner.

The best method employed in obtaining the change of direction to the right consists of the combined use of the right direct rein and the left indirect rein of opposition in front of the withers. The former displaces the head slightly to the right, while the latter prevents this from becoming exaggerated, ties the shoulders to the neck and causes them to take the same direction.

The legs must maintain the march. If the action of the reins tends to slow up the horse, the change of direction is either executed poorly or not at all. The legs should therefore be ready to act, when needed, either to cause the horse to obey the rein aids or to maintain the gait if it should tend to decrease as a result of the action of the hand. This necessity of advancing, moreover, makes it obligatory that the haunches deviate neither to the right nor left, or the propulsion of the hind legs will lose their effect. Thus the legs must maintain the hindquarters in their proper place and inclose them in order to prevent, when necessary, their lateral displacement.

If the rider, for example, after having turned to the right, wishes to resume the march to the front, he should distribute the weight of the forehand equally on the two shoulders by the equal and direct action of the two reins. Throughout this action the legs act with equal force, if necessary, to straighten the horse and to push him forward in his new equilibrium.

b. *The broken line, circle, and serpentine.*
When the turns are demanded by the reins alone acting on the forehand, they have the effect not only of suppling the shoulders but also of favoring the engagement of the haunches.

The same movements, when the leg pushes the haunches to the outside, give great mobility to the hindquarters. In using them one should, therefore, have in mind the end to be obtained and employ them accordingly.

Mobility of the hindquarters is desirable only insofar as there is strict obedience to the leg. Its ultimate object is to enable the horse to be kept straight under all circumstances.

c. *The half turn.*
The half-turn, with the radius progressively decreased, leads to a more and more marked engagement of the haunches. The half turn on the haunches (about on the haunches) is the final expression of the half turn, and demands the ultimate degree of engagement of the hocks.

d. *The half turn in reverse.*
The half turn in reverse, as the radius is decreased, tends to mobilize the haunches. The half turn (about) on the forehand, which is its limit, gives absolute mobility of the hindquarters by displacing them about the forehand.

103. TO OBTAIN THE ENGAGEMENT AND THE MOBILITY OF THE HINDQUARTERS.
The hindquarters are the seat of impulsion and at the same time they form a sort of rudder in the changes of direction.

The mechanism of impulsion lies in the play of the hip joint (coxofemoral articulation). The closing of this joint leads to the engagement of the hocks under the mass and allows the horse to cover more or less ground according to the energy of the extension of the propellers. Such engagement of the hocks under the mass leads to a lowering of the hindquarters—a position which greatly facilitates rapid changes of direction.

The hindquarters should be able to move with facility in a lateral direction as well, but because of his construction the horse can execute this movement only by passing the right hind foot, for example, in front of the left hind. Here again the horse must lower his croup and draw his hocks under the mass.

The movements which cause this engagement and this mobility of the hindquarters are: extending and collecting the gaits, halts, the broken line, the serpentine, the circle, the half turn and the half turn in reverse with a smaller and smaller radius, the false gallop [counter-canter], backing, two tracks and shoulder-in.

104. TO BACK.

Backing is a movement which should not be attempted until the training of the horse has progressed sufficiently so that he moves forward on the bit and halts calmly.

The first few lessons should be given dismounted. Standing in front of and facing the horse, grasp the reins, one in each hand, a few inches from the bit. With light half-halt effects to the rear and use of the voice, induce the horse to take one or two backward steps. Then promptly lead him out to the front and reward him.

After the horse has learned the dismounted lesson, he is ready to learn to back with the rider mounted. With the horse moving forward, he is halted in the normal manner. Just as he halts, and before he has time to set himself, the rider increases the rearward action of his reins, applying a series of half-halts. At the same time, use the legs actively, squeezing the horse slightly well forward on the girth.

This enables the rider to begin the backward movement while the horse is still unbalanced. Once the horse gives and begins to back, applying the legs and using a series of half-halts will keep him backing. Not more than two or three steps should be required. Failure to back straight should be corrected by strong use of the legs.

When the horse resists backing, a dismounted assistant may stand in front of the horse and step on his coronet; or tap him lightly on the legs with a crop, at the same time the rider applies his aids.

Backing often may be a punishment for a horse if, in spite of halts and half-halts, he seeks to force the hand or to bear heavily on the bit, but it is a means whereby the spinal column is made supple and it helps the horse to place himself on his haunches. Backing represents a further degree in the gymnastics of alternately reducing and increasing the base of support.

The suppling undergone by the young horse in the extension and collection of gaits will generally lead to his backing without difficulty. Sometimes, however, through pain or stubbornness he may refuse to back, he may brace himself with his haunches, contract the spinal column and resist the action of the reins.

These bad habits if left alone would result in a habitual reluctance to back. The rider should overcome them by displacing each of the shoulders to the corresponding haunch. He must profit by the displacement of the haunch to resume the action of the hands.

In the execution of this movement the horse should be calm and straight. He should back slowly, at the same time be ready to move forward at the call of the legs. In the beginning it is best, especially with high spirited horses, to follow the movement of backing by again moving forward, and then by rest with reins free.

When these results have been obtained, the horse should then be made to pass more freely from the forward to the backward movement, and vice versa, and thus be balanced between the two. In this exercise only a few steps should be taken in each direction.

105. TO OBTAIN FREE PLAY OF THE SHOULDERS.

The horse at liberty moves and balances himself with ease, but the young horse generally becomes heavy on the hand when mounted. This change results partly from the addition of the rider's weight, and partly because the horse can balance himself for movements which he himself wishes to execute, but does not yet know how to balance himself for those which his rider demands.

The gymnastics best qualified to give mobility to the shoulders and lightness to the forehand include: slowing; halts; half-halts; backing; the broken line and serpentine with the turns being demanded by the reins alone thus making the forehand pivot about the haunches; half-turns progressively smaller until the half turn on the haunches is reached; two tracks; and shoulder in.

Broken lines at the gallop, with continually shorter turns, are of especial advantage. This is the best training to render the horse supple and mobile in the gallop, easy to guide, and clever on varied ground.

All these movements are demanded by the bearing rein, which acts indirectly but effectively upon the shoulders.

It should be understood that the various movements enumerated above will not in themselves accomplish the desired results. The rider must keep his object constantly in mind and when resistance is met, he must act with tact; this is, with more or less energy or gentleness according to the circumstances.

The various movements should be wide at the beginning in order not to discourage the horse; later they can be made progressively smaller so that finally the complete submission to the aids—necessary in individual combat—will be obtained.

The *forward movement* must be carefully preserved during all gymnastic work, and it is always required after collected work in order to let the horse stretch himself and extend the gait before he is rested. This avoids the danger of losing impulsion.

106. SUMMARY.

The horse should now be ready for the exercises in Chapters VII and VIII and he should be practiced for further training.

107. CONDITIONING.

Conditioning of military horses is somewhat closely related to the training of race horses. The application of rules of hygiene and the consideration given to the processes of physical development are of course the same. Progression in the work is also derived from the same principles. Conditioning the military horse does not require his preparation for trials of speed on certain days, however. Rather, by the rational development of all his organs and physique, generally, he must develop, and retain throughout his service, that endurance, hardiness, and agility on varied ground which are indispensable to the cavalry mount.

Therefore, one cannot fix absolute rules for the conditioning of the young horse. His condition, appetite, the preservation of his legs, and his general appearance are the only regulations and guides, for training is an art and it evinces all the qualities that distinguish the true horseman.

The conditioning of the remount is coincident with his suppling. Logical conditioning requires daily lessons lasting between one and one-half hours as a minimum and three hours as a maximum. All instructions given in breaking which concern the general condition of the horse, i.e., the importance of proper feeding, development of muscles, and the care of the legs and feet should now be observed even more carefully, for the work during training is more intensive.

Scrupulous regularity of gaits is the basis of conditioning. The instructor assures the success of this work by employing a reasonable schedule of periods at the trot and gallop, combined with periods at the walk and at rest.

This work should continue throughout the whole year. Each week requires a slightly greater effort and a consequent development of the body. It is sometimes advisable to break the long ascent to condition by rest periods so that the horse may take hold of himself, renew his strength, energy and spirit, and prepare himself to respond to the new demands that will be made upon him. At times it is worthwhile to cut down, or even stop the work, and substitute mere walks in hand, especially in the case of horses of delicate temperament or of those that have weak legs.

A horse in good working condition presents a general appearance of health with wide-awake eye, brilliant coat, supple skin, and muscles standing out; his flank is well let down and his gaits are easy, calm, and indicative of strength.

108. USE OF ARMS.
The horse should be accustomed to arms and the handling of arms early in its training. In training to bear the saber, the wooden saber scabbard alone should first be attached to the saddle; later, when the horse bears this without fear, the saber is used. When, in the riding hall, the horse is no longer afraid when the rider draws or returns pistol or saber, the rider should carry these weapons occasionally on long rides out of doors, during which he should quietly draw and return them and execute simple movements illustrating their use. Care must be exercised not to touch or strike the horse with a weapon in a manner to frighten him.

A few minutes at odd times should be devoted to close-order drill. The horses are thus early accustomed to the pressure of ranks and to acting together in unison.

Precaution must be exercised in accustoming the horse to the sound of firing. In the beginning the pistol should be fired some distance from the horses and while the remounts are being worked in an enclosure, such as a sand ring. The firer can gradually come closer, and after a few lessons the pistol may be fired in the sand ring. When the horses show sufficient quietness, the pistol may be fired by the riders of the quietest horses.

As the training progresses the pistol may be fired at the completion of a cross-country ride.

The pistol should be fired well away from the horse's ears. The rider must remain absolutely quiet while the pistol is fired, and if the horse jumps he must not jerk his horse, but must endeavor to quiet him by taking him up smoothly and by use of the voice.

When the horses show sufficient quietness and relaxation during pistol firing they may be practiced on the pistol course.

109. PREPARATION FOR FIELD SERVICE.
  a. During the training period, horses should gradually be accustomed to carrying full field equipment and to marching.
  b. In this period, men and animals should be trained in the following:
  (1) Mounting and dismounting.
  (2) Leading—on the road, over low obstacles, and over rough ground.
  (3) Correct gaits.
  (4) Moving out and increasing gaits together.

(5) Decreasing the gait properly.
(6) Maintaining correct distances.
(7) Marching in column of fours, in line, deployments, leaving the column to go on small missions, such as couriers, flankers, advance guards, rear guards, patrols.
(8) The noise and the movements involved in the manual of arms.
(9) The noise of firing near by.
(10) Unsaddling and tying on field picket line, feeding, watering, re-saddling, mounting, and forming ranks.

c. The work should be conducted in a manner to demonstrate the marching capacity of the animals and their general fitness for field service. Any defects observed in their training should be carefully noted for correction during the remainder of the course.

CHAPTER XV
TRAINING THE HORSE TO JUMP

110. Leading 195
111. Jumping on the Longe... 196
    a. General 196
    b. Danger of longeing over broad obstacles 197
112. Amount of Jumping to be Done............................. 198
113. First Jumping Mounted... 199
114. Jumping at the Trot.. 199
115. Jumping at Liberty 200
116. Jumping at the Gallop.. 200
117. The Arrangement of Obstacles 201
118. In-and-Outs 201
119. Exercise Before Jumping 202
120. Riding the Green Horse Over the Obstacle 203
    a. Importance of leaving the horse alone................. 203
    b. The approach 204
    c. Clearing the obstacle... 205
    d. Landing 206
121. Training over Courses of Jumps 207
122. Jumping Without Wings... 207
123. Rapping.. 208

110. LEADING.

A month after breaking begins, or as soon as the colt goes well at the walk and trot on the longe, he should be led across poles or small logs laid flat on the ground. Sugar and oats fed to timid colts immediately after obstacles are crossed have a tremendous influence in creating calmness and boldness.

This applies both to the colt and to the older horse when the latter is being schooled over big and strange obstacles. Leading is not only the best way to start jumping lessons, but is often useful later when, during hunting or hacking, it is wiser to lead one's mount over some tricky little barrier than to jump it. In leading, no difficulty whatever will be found if the height is built up gradually to eight or ten inches. The colt at first should be required to *step* across and *not jump* the obstacles. In the beginning they should be low enough so that in stepping over them *he does not bump his legs.* The ideas in mind are to give him confidence and to teach him to lead over unusual places, and it is well worth the trainer's time. Later, after learning to jump on the longe, leading over obstacles high enough to require jumping occasionally should be practiced.

In this leading, the obstacles, wherever possible, are rigidly fixed; otherwise heavy poles, hexagonal or octagonal in shape and having sharp edges which will hurt the colt if struck, are used. This rule applies to all future training in jumping. Hedges, or obstacles easily knocked down, teach carelessness and disrespect. Small ditches, from one foot to two and one-half feet wide, are led across concurrently with the instruction over upright obstacles.

111. JUMPING ON THE LONGE.
a. General.

Generally after about six week's schooling, but only when the colt has perfect manners at the walk and trot, his instruction over obstacles while circling on the longe may begin. As always, obstacles are practically rigid, and the end away from the trainer is enclosed by a high wing. In all early instruction in jumping, every advantageous device which may prevent refusing, running out, or becoming excited, ought to be utilized. In other words, if the trainer takes ample time and uses patience, ingenuity and common sense, bad habits need never be acquired by the horse. For this reason, an assistant should be on hand when first lessons over obstacles on the longe are given. He follows the colt from in rear, and by use of voice and threats or light touches with the longeing whip, prevents hesitation and refusals. Whenever any excitement or apprehension appears the obstacle should be lowered. Above all, both on the longe and later when mounted, calmness and the same gait should be maintained. Loss of calmness, in any training, proves lack of sufficient preparation.

When longeing over an obstacle, the trainer moves up and down the long axis of an ellipse, which permits the colt to approach and leave the obstacle on a straight line. This is important, for in early training he should not be required to jump while moving on a curve. From the bar on the ground, the height is increased inch by inch, but with the colt still being required to approach at the walk. From experience in leading, he normally will continue to step over, and not jump the fence. However, when the height has been increased to twelve inches or more, occasionally he will bump his legs or stumble, in stepping over the bar, narrow box, "chicken coop," or other obstacle. If these are rigid, or heavy, and sharp-edged, sooner or later the colt, while continuing to move at the walk, in order to avoid the painful bumps will decide to jump. Every effort is made to keep him moving calmly at the walk *after jumping* as well as while approaching. As soon as he jumps while walking without excitement, the height may be progressively increased up to two feet, or possibly higher, depending on the manners and natural ability of the individual. The increase should be made very slowly; one or two inches every three or four days. This careful preparatory training requires patience, but as stated before, it will prove most profitable.

The jumping on the longe should take place at the end of the day's work when the colt's "edge" is worn off. If he is fresh he inevitably will "play" and discover undesirable tricks. Great care to avoid overdoing the practice and generous rewards for good conduct soon produce adroitness and calmness. The time spent jumping while going to the left and the right hand on the circle should be equalized. Constant variation in type, appearance, height and eventually breadth of obstacles employed is most desirable. *However, the use of a take-off bar on the ground to increase breadth is not recommended.* It is unnatural and requires so much study during the approach that it interferes with the development of good form. An efficacious device for making a horse "stand back" at his take-off will be described later on. In-and-outs should be employed as soon as calmness and skill permit.

Well-conducted lessons in jumping on the longe are the quickest and most efficient means of making a calm jumper. This implies that the horse is well disciplined and obedient on the longe before any jumping is undertaken.

b. Danger of Longeing over Broad Obstacles.—*In connection with breadth, it is very dangerous, even with experienced horses of great jumping ability, to longe over exceedingly broad obstacles.* Breadth and unusualness of appearance often may puzzle or intimidate a horse as he studies the obstacle during his approach and cause him to decrease his speed. If his slowing of rate is sudden and marked, it is practically impossible to prevent it with voice and whip. He then either may refuse, or, because of insufficient impulsion to carry him over, fall into the obstacle. Even though he receives no injury, *falling into an obstacle* generally brings loss of courage for a long period of time even to a very bold horse. Striking an obstacle, but falling clear of it, never has the same *demoralizing effect* which follows a fall into it. The trainer, when mounted, by use of his spurs when approaching a very broad obstacle, usually can compel sufficient impulsion to clear it.

112. AMOUNT OF JUMPING TO BE DONE.

In jumping, more than in any other phase of training, horsemen are prone to proceed much too rapidly, thereby pushing the colt beyond his ability. The latter then loses his confidence, courage and form. To restore them all is usually impossible. Always the thought should be kept in mind that much practice over jumps well within the horse's ability finally produces that prime requisite—good form. As good form is slowly and carefully perfected, the jumps may be broadened and heightened inch by inch and a fearless jumper developed. He then will stand well back on his take-off; jump without decreasing his speed or putting in short, pecking strides; describe a long, graceful parabola of flight over the obstacle and land smoothly "going away." The moment any timidity, nervousness, rushing or lack of form is evinced, training over smaller jumps should be resumed.

When training a *green* hunter or jumper, young or old, practice two or three times a week over ten or fifteen moderate obstacles is ample. With an *experienced* old one, wise horsemen do very little jumping until two or three weeks prior to the show, or other event, in which the horse is to participate. At other times, work over *five or six obstacles once a week* is enough schooling for such a horse, and none of these should be up to the height the horse is capable of negotiating. Of course, when preparing him for competition, one or two of the obstacles should be a few inches higher and broader than those to be encountered, but when he jumps them well *once*, it is time to stop work. If he does the whole course well except one or two obstacles, he should be jumped over those he failed on and then sent home. If the size frightens him, the obstacle should be lowered and built up gradually along with his confidence. Unfortunately, this routine usually is not followed. Most hunters and jumpers are ruined by too much too-high jumping, not by too little.

Normally the four-year-old green colt should not be asked to jump higher than three feet while mounted, and not over four feet on the longe. *The principal test, in order to determine whether any horse is fit to jump on a particular day (after his legs and feet have been inspected by sight and touch for enlargements and fever) is accomplished by trotting him on hard, level ground. If any sign of stiffness or lameness exists, under no circumstances should he be required to jump.*

113. FIRST JUMPING MOUNTED.

After about two week's jumping instruction on the longe, mounted work over the same obstacles that were employed for leading longeing may begin. When mounted, the trainer, beginning at the walk, rehearses thoroughly all work previously done on the longe. This is done despite the fact that the colt, when on the longe, by this time may be cantering over small obstacles about two feet high. The points of most concern, when mounted work starts, are: allowing absolute liberty to head and neck; maintaining the same speed and gait before and after passing an obstacle; and approaching the center of each obstacle perpendicularly. The rider should hold with one hand either to the pommel, or to a strap around the horse's neck, in order to prevent being displaced and so alarming and hurting the colt through jerking his mouth or jolting his back. *The reins, without exception, should be loose and floating when the colt actually crosses the obstacle.* In fact, the reins should be floating *at all times in these early lessons.*

If a certain horse, either when mounted or on the longe, does not learn to jump smoothly *at the walk* after the obstacle is set at a height of fifteen or more inches, but instead persists in hopping over, first with his forehand and then with his hindquarters, he should be put over it at the trot. The speed of this gait will make it necessary for him to jump, rather than to hop one end at a time. While this elementary mounted work at the walk takes place, longeing continues over higher obstacles. In other words, training on the longe and at liberty is always more advanced and difficult than that which is carried on concurrently while mounted. No increase in the height of obstacles, used either in the work executed on the longe or while mounted, should be attempted until perfect calmness exists in the work at hand. Calmness is demonstrated by the colt's approaching and leaving the jump on a straight line and with no increase or decrease in speed. Thus work at the trot, both mounted and on the longe, begins only when absolute calmness exists at the walk. Consequently, the time to commence jumping at the trot varies in each case. The mounted colt often should be very gently halted while approaching, or just after clearing an obstacle, then patted and rested.

114. JUMPING AT THE TROT.

Jumping at the trot is very beneficial for it *necessitates entire relaxation of the spinal column* and teaches clever use of the hindquarters in propelling the horse over the obstacle. Here too, the pommel or neck strap always should be held in order to avoid falling out of balance to the rear with the evil results mentioned above. At the trot this is very necessary, for the reaction of the horse's jump on the rider is violent. Jumping, while at the trot, over all types of obstacles, which have been gradually heightened to about two and one-half feet, should continue until perfect manners are habitual. If excitement develops, jumping at the walk is resumed. Only after all goes perfectly at the trot does the trainer attempt jumping at the gallop. In leaving the obstacle the horse is not permitted to turn continually in one direction. The trainer's imagination ought to keep the pupil expectantly attentive and allow him no initiative.

115. JUMPING AT LIBERTY.
Work over obstacles at liberty is excellent practice for both colts and older horses, provided an elliptical pen is used. The great disadvantage of a straight jumping chute lies in the fact that colts almost invariably develop the bad habit of rushing their jumps. With the elliptical pen, the rather sharp turns at the ends not only prevent rushing but also improve the horse's balance. In it, every type of obstacle may be introduced to the horse. Moreover, much experience may be gained with far less chance of leg injuries than when the horse is handicapped by the rider's weight. *For young colts, jumping at liberty should be postponed until calmness over small obstacles at the walk and trot, both on the longe and mounted, is obtained.* To teach jumping at those slow gaits before taking a single obstacle at the gallop, normally will eliminate any tendency to rush. An able horse so instructed, and capably handled, ultimately will be exceedingly clever and brilliant because no excitement will interfere with his cool, experienced judgment.

As soon as conduct at the trot makes readiness for jumping at the canter evident, work in a Hitchcock or elliptical pen may be advantageously begun.

116. JUMPING AT THE GALLOP.
Normally, three or four months from the time breaking begins, the colt's strength and balance warrant jumping at the canter on the longe. Therefore, at the end of five or six months the colt may be expected, with the trainer up, to jump in good form at the canter. The obstacles ought not to exceed two feet in height. It is wisest to limit the height for the mounted work of four-year-olds to three feet. Many will boast that their horses at that age have cleared five feet. However, the advice just given holds good if one has only one or two colts to expend and desires sound six-year-olds.

The trainer's weight is a vital factor in early jumping. If much over one hundred fifty pounds, great caution must be taken to avoid injury to the colt's legs, and the greater part of his jumping should be done on the longe.

The heavy trainer, however, may longe his colt over obstacles with a light assistant riding. This, incidentally, is a sure way, if the assistant is given no reins, of preventing harm to the mouth.

117. THE ARRANGEMENT OF OBSTACLES.

From the day leading over the poles laid flat on the ground begins, the arrangement, color, shape, type and spacing of obstacles constantly should be varied. With a little imagination this can be done with a very limited number of obstacles available by using blankets, ropes, chairs, buckets, hay bales, branches, etc. The spacing and arrangement may be infinitely diversified. This training, at first thought, may appear to be more essential for a show jumper than a hunter. Nevertheless the latter will be a far safer, bolder and more clever horse in the field if he receives a similar education.

118. IN-AND-OUTS.

In-and-outs, with frequent changes in spacing and type, furnish the most effective means of developing agility, judgment and balance, over and between jumps. Again it is pointed out that this training—being a means, not an end—is just as important for the hunter as for the show jumper. Starting with two obstacles, the colt is schooled, as time goes on, over triple, quadruple, and quintuple in-and-outs.

At the walk, the spacing between jumps for the green colt may be any distance from five yards up. At this gait, the obstacles are never over two and one-half feet in height and the colt easily alters his length of step so as to take off correctly. At the trot, the usual spacing is six yards; at the gallop, *with low obstacles not over three feet in height,* seven yards; and for obstacles four feet or higher at the gallop, eight yards. When, at five or six years of age, the horse becomes skillful, these distances may be changed, as for instance by having only seven yards between two high obstacles, or four yards between two fairly low ones, which will require the horse to "pop" in and out without an intervening stride at the gallop. *As always, it is much practice over* low *in-and-outs which gives calmness and agility and ultimately makes a great jumper over any type of jump. Patience brings calmness, and calmness permits cleverness.* The distracted horse which fears and hates jumping can never be a consistent performer. Sad to relate, not over one out of ten hunters or jumpers receives thorough and gradual elementary education in jumping. As an example of a training in-and-out for a four-year-old after five months' mounted work, a single bar two feet high may be followed at six yards by a small double fence two and one half feet high, with the two fences two feet apart; this, succeeded at eight yards' distance by a two and one-half foot wall. Later, five or six small obstacles may be arranged at varying intervals within a chute formed by wings or fences. When mounted, the objections found to jumping in a straight chute while at liberty are not present. The triple in-and-out described above may be taken at the walk, trot, or gallop. The colt frequently should be halted between two of the obstacles, patted, turned about and jumped out over the one over which he entered. Or, he may be turned ninety degrees and taken out over a small barrier on one side of the chute. Obviously the arrangements may be varied infinitely and, as conditions allow, a slight change in direction after each jump is introduced. By the end of five month's training, gymnastic exercises will have made the colt supple and responsive to the legs and reins so that the halts and changes of direction during practice at jumping can be executed smoothly and calmly.

A colt, as he gains experience, may be longed over simple and triple in-and-outs. The side of the obstacles away from the trainer must be walled, or fenced in by wings. The uprights toward him must have poles—with one end resting on the ground, the other on the top of the upright—so that the longe will slip over and not catch on the uprights. Here again, instruction begins at the walk. Punishment on the nose with the longe will prevent rushing, and since in-and-outs by their nature discourage it, the colt can be taught to go calmly. The trainer will find it necessary to "double time" parallel to the horse's croup as he longes him through an in-and-out, both because the longe is usually not long enough to permit his standing still, and because he must prevent the colt's stopping.

In addition to the in-and-outs, single jumps should be taken, preferably after work over the former, and when the horse is relaxed and calm. The single jumps are a little higher and broader than those comprising the in-and-outs. By jumping these, the colt gains confidence in his ability over larger obstacles, while his form is perfected over the small ones in the in-and-outs.

119. EXERCISE BEFORE JUMPING.

As a fixed rule, no jumping ever is done *during training* until the horse has been worked sufficiently at hacking, shoulder-in, changes of rate and gait, circles, serpentines, broken lines and other exercises—*according to his particular needs*—to wear off his first playfulness. He then will be calm, relaxed, obedient and softly on the hand.

In this connection, when work over obstacles *at the gallop* is initiated, the rider begins to maintain soft contact with the mouth and ceases to hold the pommel, as is done in the elementary lessons at the walk and trot. No matter what occurs, the fingers always ought to be semi-relaxed at an obstacle, so that in case of a "bad jump" the reins may slip through them and the horse's mouth not be hurt.

With a finished jumper, the "higher" he feels before a jumping competition, the better are his prospects of winning. He has no need of all the calming exercises that a green horse must have prior to *training in jumping*, and the absurd habit of "working the horse down" before he enters the show ring simply take a few potential inches off each leap. Of course, if he is old and stiff he must be "limbered up," but if not, a bit of cantering and a couple of jumps to warm his blood are all he needs. For a week before a competition the finished jumper requires only the lightest sort of work with almost no jumping. This permits him to store up an abnormal amount of energy for the crucial moments when big, strange obstacles confront him. Likewise a well trained hunter needs little schooling over obstacles, for he receives all the jumping necessary to keep him in form while hunting. On the contrary, if a horse is *not a trained jumper*, he is not ready to show or hunt, for the show ring and hunting field are not suitable places for him to learn the mechanics of jumping. He and his rider are nuisances to others, and the nerve-wracking conditions prevailing generally do the horse's disposition and jumping form more harm than good.

120. RIDING THE GREEN HORSE OVER OBSTACLES.
a. Importance of leaving the horse alone.
The main and most difficult task of the trainer when riding over an obstacle is to "let the horse alone." Simple though this sounds, in practice it requires cool nerves and great coordination. Riding a jumper is a specialized form of athletics. It may be very well done by riders having little general ability in equitation or horse training, provided they have good natural balance and muscular control. Obviously the trainers of hunters and jumpers need perfect form in jumping to accomplish properly this phase of training.

Too often riders believe that they are assisting their horses over obstacles by using their hands and legs in various ways other than those to be described. The idea that one can "place" his horse for each jump over a course of big and imposing obstacles is erroneous. Many really brilliant riders have tried it, but without complete success. The horse must do the jumping and the less he is bothered, except to encourage and rate him, the better he will do. If "placed" for each jump, either his or the rider's nerve or judgment inevitably cracks at some point while traversing a difficult course. Under strain, few horses will subordinate *completely* their will and actions to the men on their backs. *With the best-trained horses the tactful horseman sagaciously compromises when blood is hot and excitement rife.*

b. The approach.

Approach the center of the obstacle perpendicularly. Sit in balance with loin hollowed out and body inclined well forward from the hip joints. Do not stand in stirrups during the immediate approach, but allow your weight to settle on your thighs by relaxing the knee joints. Keep head and chest lifted.

If, as good jumpers often do, your horse starts with a rush at the first obstacle, in preparation, from the hips, lean a trifle farther forward than usual in order not to be "caught behind." The momentum of the quick start will throw your body upward and backward into the correct angle. If this is not done, you may be caught out of balance and forced to hang on the reins.

Do not lean forward or backward in anticipation or apprehension, just prior to the take-off. Sit still.

Keep the reins *very slightly stretched* and be certain that the hands, with relaxed fingers, *accentuate* the following of every oscillation of the horse's head and neck, beginning with his first stride in the direction of the obstacle. Pulling during the first few strides, as he gains impetus, will make a horse nervous, and eventually develop a frantic rusher. His "balancer" must have the absolute freedom that results from complete relaxation of finger, elbow and shoulder joints.

If the reins are barely taut, the slightest resistance by the fingers will collect and rate the horse during the approach (if that is necessary) *for the horse on a light rein is relaxed in poll and neck.* If, on the other hand, you are resisting or pulling with the reins, he stiffens and resists so that your efforts have little effect in rating or changing direction.

While approaching, your legs are continuously active, with the intensity of action varying according to the sensibility of the horse, but always sufficient to sustain the speed required by the height and breadth of the obstacle. The best jumper in the world some day will refuse if the rider's legs are passive or weak. The legs' action may be intermittent and in rhythm with the strides, being performed by squeezing with the calves or tapping, when necessary, with the heels or spurs. During the last few strides before the obstacle, determined squeezing is usually sufficient encouragement. With legs tight to the horse the seat is secure and the spurs may be turned against his sides at the last instant if you sense a refusal. With experience, your hands and seat learn to anticipate any attempt at disobedience.

Combine the indications of heart, balance and legs to encourage the horse to gallop in his stride boldly over the obstacle. Indecision on the part of your mind, legs or position—which is usually abetted by apprehensive clutching of the reins—quickly transmits the same indecision to the horse and causes a run-out, a refusal, or a "pecking," timid jump.

Pinch more tightly than usual with the knees during the approach, in order to prevent slipping forward in the saddle, while the lower legs are engaged in driving the horse onward.

Be tactful in convincing your horse that he *must* jump, *for too much activity of legs and failure to sit quietly* before the take-off communicate similar excitement and nervousness to the horse with consequent loss of calmness and cleverness. He must know that you intend he shall jump, but he must not believe from your actions that you are excited about the matter.

c. Clearing the Obstacle.

As the horse jumps, hold your forward inclination (with back straight, loin hollowed out, head and chest lifted) so that the whole body, from the knees up, is projected out of the saddle as a result of its inertia when the horse's forehand checks and rises. At this moment stiffen the knee joints as they inevitability open part way, and allow your entire weight to sink into your depressed heels. Your knees remain pinched against the saddle. Thereafter, while clearing the obstacle, you remain in balance out of the saddle, supported by your knees and stirrups. The ankles, knees and hips are the springs and joints which open and close as necessary to maintain balance and to soften the shocks of the take-off and landing. Force your heels well down before starting for the obstacle so that your feet are braced against the stirrups. This position of the heels automatically gives a strong grip with the calves, and helps balance immeasurably, both over the obstacle and upon landing.

Contact with the mouth, which by degrees should become feathery light during the approach, vanishes at the instant the horse rises over the obstacle to the mere weight of the reins. With any save a very expert horseman, it is far better to permit a floating rein during the period of suspension, for the slightest resistance, pull, or jerk against a delicate mouth prevents full use of the balancer and maximum folding of the knees and fetlocks. Consequently a fault by the fore legs is almost certain. In addition, a few blows to the mouth will dissipate calmness and boldness.

Over the jump, the horse pivots between your fixed knees, and although you feel your lower legs fairly snug against his sides, the stirrup straps necessarily move a trifle in rear of the perpendicular while he is ascending. As momentum dies out over the top of the jump, the stirrup straps become approximately perpendicular. In other words, your lower legs feel to be, but in reality are not, fixed to the horse's sides. The knees however remain fixed in position.

The habit, so irritating to the horse, of allowing the lower legs to slide far to the rear along his sides, is unsightly, upsetting to the horse, and destructive to security of seat. On the other hand, lower legs in the correct position allow you to straighten up your body from the brace of your feet against the stirrups, in case of a bad landing, and so prevent your being pitched forward on the horse's neck or over his head, as is often the case where the legs slide to the rear and the toes point downward. The same conditions are applicable to refusals.

d. Landing.

Upon landing, do not commit the vicious mistake of roughly snatching the reins, thereby terrifying the horse and hurting his mouth. He, if intelligent, can only interpret such treatment as punishment for jumping and soon will begin to refuse. Reestablish normal contact with his mouth very gently during the first strides after landing.

As the horse descends, let your lower legs slip a trifle forward so that the stirrup straps are the least bit in front of the vertical. This position is favorable to receiving your weight in your heels and so maintaining balance while remaining out of the saddle upon landing. Your ankle, knee and hip joints absorb the shock, but remain elastically stiff so as to keep the seat entirely off the saddle.

Provided there is another obstacle close at hand to be jumped, relax your knees gradually two or three strides after landing, and sink into the saddle, your weight supported on the thighs and crotch.

If a turn in direction immediately after jumping the obstacle is obligatory, it is initiated by very softly using the opening rein while your horse is in the air over the obstacle. The lightest effect suffices and the opposite rein is allowed to be passive. Abrupt changes of direction should not be required until the horse has become a well trained and confident performer.

To prevent the colt's running out, and also when it is necessary to change direction just before or after a jump, the opening, or direct, rein is normally used. Neither the direct nor the indirect reins of opposition should be used when the opening or bearing reins will accomplish the desired result. The latter are conducive to calmness, while the former distract the horse when he is jumping.

121. TRAINING OVER COURSES OF JUMPS.

When the colt becomes calm and proficient in jumping at the walk, trot and gallop, he should thereafter receive his principal jumping practice by galloping over progressively longer courses of obstacles. The courses are varied at each lesson as to succession, type, number, spacing and arrangement of obstacles. The sharpness of changes in direction and the length of the courses are increased very gradually.

The exercises in extending and slowing the gait which are being concurrently carried on, by this time should have made it possible to rate the horse readily.

When executed at the gallop, the trainer endeavors to maintain a free, long-striding, uniform gait and encourages the horse to jump in his stride. If the colt shortens his stride excessively, or puts in extra strides before each jump, it generally indicates that jumping is painful to some part of his anatomy. Often his feet or muscles may be a little sore although no lameness is present. Rest is essential or bad habits and poor form will result.

The most pleasant characteristic of an excellent hunter or jumper is his willingness to be "rated." If trained for a year or more according to preceding chapters before he is called upon to hunt a difficult country, or to jump a full-sized course of imposing obstacles, rating will be possible. After all is said and done, the rider's main task when hunting and jumping is to establish the proper rate of speed according to the type and size of obstacles facing him; from then on, the horse handles the task. The numerous frenzied and uncontrollable horses seen in the hunting field and show ring are a sad commentary on their trainers. Riders of varying degrees of competence often attempt to conduct over obstacles horses which, in their untrained state, could not be ridden at a uniform gait around a circle in a riding hall by a most skillful horseman. In most cases these demonstrations are directly attributable to improper preparation.

122. JUMPING WITHOUT WINGS.

As the horse becomes perfectly amenable to the various rein effects, during some jumping lessons wings should be eliminated. Again common sense dictates that work without wings first be done over the elementary obstacles with which the colt has long been familiar. Using a very soft vibration in anticipation of a runout is a wise preventive measure. This is also applicable to an old horse when he faces a strange fence. It is a foolish risk, when avoidable, to attempt to put any horse over a large and strange obstacle for the first time if it is not framed by wings. Intelligent care may prevent a good horse's refusing or running out throughout his entire career. The trainer must recognize each colt's potential ability and never ask more than he can comfortably accomplish. Certain it is that many horses never can be outstanding jumpers. Great ability and boldness must be innate in truly famous performers. However, much can be done with little, where patience and intelligence serve proper methods. Ultimately, with proper training, a horse will have no need of wings because he knows that "crime doesn't pay."

123. RAPPING.
If correctly used with careless or sluggish horses, rapping bars are beneficial. They should be "rapped" over low jumps, about two and one-half feet high, with a rapping bar handled by two men. When poorly manipulated—as is unavoidably the case with a single assistant—the bar can be seen by the horse, and his fear soon causes him to peck or refuse. Each of the two men holds one end of the bar which, from its concealed position back of the top element of the obstacle, is raised several inches just as the horse is actually clearing with his fore legs. In order not to frighten him the bar is raised quietly and not very high so that several jumps may be made before it is hit by the horse's front cannons. If, on the following jump, the horse adds several inches to the height of his leap, the rapping has accomplished its purpose; if not, the procedure is repeated. A broomstick may be used to tap the hind legs when necessary. Usually faults caused by dragging the hind feet result from heavy hands or a backward seat, which prevents the horse using his head and neck. A bar of hollow iron which is not exceedingly heavy but which will not bend is most satisfactory.

If severely punished once or twice by rapping, a careless colt, for a long time afterward, usually will be much more attentive. Likewise an experienced but sluggish hunter may be made a safer conveyance by a little judicious rapping. This method of punishment has no place in the early training of a colt, and in general the use of rigid schooling jumps is the most satisfactory means of instilling respect for all obstacles. For a horse so trained, rapping will be unnecessary.

To those who object to rapping on the grounds of cruelty to animals, the answer is: it is better to make the horse's shins smart now and then than to let him break his own or his rider's bones.

CHAPTER XVI
TRAINING THE RACE HORSE

124. The Seat 209
    a. Flat racing 209
    b. Steeplechasing 210
125. Training and Conditioning Periods 210
    a. General 210
    b. Health building period 210
    c. Muscle building period 211
    d. Wind developing period 212
126. Training the Steeplechaser to Jump 212
    a. Jumping on the Longe and at Liberty 212
    b. Jumping mounted 212
127. Conditioning Schedules 213
128. Cautions in Training 216
129. Suggestions on Riding the Race 216

124. THE SEAT.

a. Flat racing.

This is a special form of riding. The speed element demands that, for maximum results, the horse have great liberty of body movement and an advanced center of gravity. The rider consequently shortens his stirrups considerably, rises out of his saddle and supports himself almost entirely with the stirrups, knees and legs. He advances and lowers his center of gravity over this shortened base of support by an extreme forward inclination of upper body and flexion of the knee, his back maintaining its normal posture. This position over a shortened base of support (due to buttocks being out of the saddle but with the crotch still near it) entails lack of stability and renders maintenance of balance more difficult, but it ensures the horse greater freedom in his gallop and advances and lowers the center of gravity of the rider, bringing it nearer to that of the horse—a condition favoring speed.

b. Steeplechasing.
In steeplechasing the rider seeks to combine speed with security. His position is less extreme than in flat racing. It is determined only after much experience and depends upon the individual, the horse and course to be negotiated. Ability to use the legs, and security over jumps, are necessary. Therefore, as compared to the racing seat, speed requirements of the rider's position are disregarded just sufficiently to attain this end. The rider's base of support is lengthened by closer contact of thighs, crotch, and at times, extending rearward even as far as the points of the buttocks. Forward inclination of the upper body consequently is not so extreme. Security is usually provided by slightly longer stirrups than in flat racing, permitting better leg grip, but the stirrups are shorter and the upper body inclined farther to the front than in the military seat. This rider can rise easily from his saddle and support his weight on stirrups and knees while galloping on the flat, and can also *"sit close"* as an obstacle is approached, to be with his horse, to feel him better and to have greater security.

125. TRAINING AND CONDITIONING PERIODS.
a. General.
The purpose of training is to enable the horse to run his race at his maximum speed, negotiating the jumps with precision and safety, while suffering the minimum fatigue. This objective should always be kept in mind. It is assumed that a horse being put in training for steeplechasing has previously had the normal training of a riding horse.

The course of training is divided into three periods: the health building, muscle building and wind developing periods.

b. Health Building Period.
In this period it is essential that the horse be brought into good flesh and carrying some excess fat. This additional flesh will become essential during the muscle building period. During this period the horse should be given only such work as will promote his appetite and keep his blood from becoming overheated.

The exercise during this period should consist of long walks, alternated with short trot periods, for one to one and one half hours daily. As the horse comes into condition the length of the trot periods should be increased. Every other day the horse may be led instead of being ridden. There should be no galloping during this period.

The normal grain ration is sufficient during this period but the horse should be given all the hay he will consume, and water should be available in his stall.

c. Muscle Building Period.

During this phase the excess flesh stored up in the first period must be converted into muscle. There should be but slight loss of weight in the process of changing excess flesh to muscle, and this must be carefully checked. The exercise during this period becomes more exacting, the trot periods are increased and short periods at the gallop are added.

At first this galloping should be conducted at a slow gallop, about twelve miles per hour. Later this should be increased to about eighteen to twenty miles an hour, but during this period no galloping at the racing pace should be attempted.

At this time it is necessary to fix the horse's galloping stride. The horse must develop a smooth effortless stride with his head low and extended. He must be taught to respond to his rider's will, increasing and decreasing his pace as demanded. If the horse is in training for a Point to Point race, he must at this time become sure and safe over difficult going and be manageable at rather sharp turns. As far as possible all this work must be carried on out-of-doors.

It is preferable to gallop in company, one horse setting the pace and the others galloping abreast of him. A bad puller, however, should be worked alone until the habit is broken.

As the work increases the grain ration must also be increased. Many horses will assimilate and require as much as sixteen pounds a day. Hay in generous quantities should be continued. A bran mash once a week, the evening before the rest day is desirable. Whenever the horse is not worked during a period of one day or more, the grain ration must be cut at least in half. This is most important as horses in high condition are subject to many internal disorders when overfed and under-exercised.

The animal's health and especially the condition of the legs must be carefully watched. Loss of appetite often indicates that the horse is being over-worked. All conditions of soreness or puffy condition of the legs must be carefully treated, and these cases should be given no work except a short period of exercise at the walk.

d. Wind developing period.

The final stage of training is the development of wind. This is best accomplished by fast gallops or, as it is commonly spoken of, "breezing."

This work must be undertaken cautiously and only after the horse has attained excellent muscular condition. At first these fast gallops should not exceed one quarter of a mile, and they should then be progressively increased; but the distance should never exceed seven-eighths of a mile.

In all the periods, but especially the last, the work should not overtax the horse; he should be pulled up before he has slowed of his own accord. Two, or at the most three, fast gallops a week is sufficient. *On these days no other work should be undertaken.*

The feeding is continued as in the previous period, except that during this period, when a horse is to be breezed, (and also before a race) the horse should be tied in his stall without access to grain or hay, and with not more than a half bucket of water, for a period of seven hours previous to the breezing exercise, (or the race).

126. TRAINING THE STEEPLECHASER TO JUMP.

The object is to train the horse to jump safely and surely at speed. The horse must be made to respect his jumps, but not to fear them.

Jumping should be conducted on the longe, or at liberty, during the health building period, and mounted during the other two. The horse's jumping education, however, should be practically completed by the end of the muscle building period.

a. Jumping on the longe and at liberty.

At the start the obstacles should be small but solid. This permits the horse to get himself out of trouble without injuring himself or losing confidence. As the training progresses, increase the size and variety of the obstacles, but remain well within the capabilities of the horse. The horse should be generously rewarded for jumps well taken, in an endeavor to develop in the horse a genuine liking for the work.

b. Jumping Mounted.

As in the work on the longe, the horse should be started over small solid obstacles from two and a half to three feet in height, with a spread of three to four feet. The use of a take-off jump in front of each obstacle, with the horse being presented at a trot, will teach respect, aid the horse in measuring his stride, and prevent rushing. As the horse acquires confidence he should be allowed to break naturally into a canter a few strides in front of his jump. As soon as the horse becomes sure and calm over the smaller obstacles they should gradually be increased in size. The type of jumps should be varied, and no longer need they be solid. The horse should now be asked to extend his gallop stride, and to jump at the eighteen to twenty mile gallop. It is important that the horse be presented straight at his fences, and that he travel straight away from them. If the horse takes hold of the bit a few strides in front of the jumps, the rider should sit still, follow his mouth with a supporting hand and let him jump at his own pace, rather than restrain him too much. The horse should not be permitted to continually get too close under his fences. If this occurs, the horse should be put back to work over smaller jumps until he learns to jump in his stride.

127. CONDITIONING SCHEDULE.
The following work schedule for a training season has produced excellent results: However it should not be followed blindly:

a. Health Building Period:

MONDAY AND THURSDAY:

1 to 1½ hours walk, with short periods at trot.

TUESDAY AND FRIDAY:

Same as Monday and Thursday, except horse is led.

WEDNESDAY AND SATURDAY:

1 hour walk and trot. ½ hour jumping on longe or at liberty.

SUNDAY:

Rest.

b. Muscle Building Period:

MONDAY AND THURSDAY:

1½ hours walk with short periods at trot and slow gallop.

½ hour jumping mounted.

TUESDAY AND FRIDAY:

2 hours walk with short periods at the "half speed" gallop.

WEDNESDAY AND SATURDAY:

1½ hours walk with short periods at the trot. Work over varied ground.

SUNDAY:

Rest.

c. Wind Developing Period:

MONDAY AND THURSDAY:

Fast gallop (breezing) ¼ mile to ½ mile.

TUESDAY AND FRIDAY:

1½ hours walk.

WEDNESDAY:

Two periods of "slow gallop" from ½ to ¾ mile. A few jumps at speed.

SATURDAY:

Half speed gallop 1 to 2 miles.

SUNDAY:

Rest.

d. General.

It is not possible to allot definite lengths of time to each of the above training periods.

The health building period is dependent on many factors, such as:

(1) Condition of the horse at the beginning of training.

    (a) In ordinary work.

    (b) In hunting condition.

    (c) Direct from pasture, where the horse has received no grain, etc.

(2) Age and previous training of the horse.

(3) Type of ration available.

Therefore the decision on the length of this period must be made by the officer in charge of training, based on his personal observation of the horses in training.

The muscle building period is slightly more definite, but again must be based on the age, previous condition and training of the horse, as well as the type of terrain available for exercise. A month should be the minimum, while three months should be sufficient for any horse selected for this type of training. However, here again the officer in charge of training must make his decision based on his personal observations.

The wind developing period is again more definite, but it also is based on the individual requirements of the various horses. It is also dependent on the type of terrain available for exercise during the muscle building period, as work through deep sand, and up and down steep slopes, during this period helps develop wind as well as muscle. Altitude and climatic conditions play an important part in the wind developing period. One month as a minimum and two months as a maximum should care for the average horse.

As stated before, this decision must rest with the officer in charge of training, based on his personal observations and experience.

The following schedule for the nine weeks prior to a race has been found very satisfactory. Depending upon the horse's condition at the start, it should not be followed blindly. Some horses in good condition could omit the first two or three weeks, others would need them.

1ST WEEK

Monday—School—work slowly 1 hour to 1½ hours.
Tuesday—Jump course or school 1 hour.
Wednesday—Gallop 1 mile, 15 miles per hour—walk or lead one hour.
Thursday—Work slowly 2 hours.
Friday—Jump course or school 1 hour.
Saturday—Gallop 15 miles per hour 1 mile—walk one hour.
Sunday—Rest.

2ND WEEK

Same as 1st week.
Saturday—Gallop 1 mile at 20 miles per hour.

3RD WEEK

Same as 1st week.
Gallop ⅛ mile at full speed. (Breeze) on Saturday after the 1 mile gallop.

4TH WEEK

Same as 1st week. Let the gallop on Wednesday be a long
cross country gallop of at least 4 miles.
Gallop ¼ mile at full speed (breeze) on Saturday after the 1 mile gallop.

5TH WEEK

Monday—Gallop 1 mile at 20 miles per hour—breeze ¼ mile.
Tuesday—Work slowly—school at slow gaits 1 hour to 1½ hours.
Wednesday—Jump course—or school.
Thursday—2 canters of 1 mile each at 20 miles per hour
(Breeze ¼ mile if wind is not good yet).
Friday—Work very slowly—school or have led.
Saturday—Jump course or gallop 12 miles per hour for 2 miles.
Sunday—Rest.

6TH WEEK
Same as 5th week.
On Saturday take a long cross country ride. Gallop at least 4 miles at about 20 M.P.H.

7TH WEEK
Same as 5th week.
Increase the breeze on Thursday to ½ mile if the horse needs more wind.

8TH WEEK
Plan on having your horse's wind and muscles at a maximum at the end of this week or just about one week before the test. Any strenuous work done the last 5 or 6 days before the test will take something out of the horse that won't come back before the test. If the horse hasn't wind enough a week before the test, he never will have it. The work this week should follow approximately the work of the 5th week.

A week before the test the horse should get a long cross country gallop to really open his lungs. After that the galloping should be only short breezes of about ¼ mile with the last breeze two days before the test.

128. CAUTIONS IN TRAINING.
Do not overtrain the horse. The Steeplechaser must be robust, and must have abundant reserve energy.

Do not let the horse sour on jumping. He must enjoy this part of his work.

Do not, in practice, require the horse to gallop, at his racing speed, the entire distance he will be required to race.

The jumps should be considered as mere incidents in a race; they should be taken in stride. The position of the rider in the field, the pace and the condition of the horse, as concerns his reserve power, should be the principal consideration of a rider during the course of a race.

129. SUGGESTIONS IN RIDING THE RACES:
Walk over the course and take note of any heavy stretches, bad going, turnings, flags, etc., and decide on the course you are going to take. Saddle your horse very carefully and in plenty of time. Always start with a definite plan based on your knowledge of your own and the other horses. Plan on where you in tend to "make your bid" and stick to it.

Practice alone can give the correct tactics to be employed in every varying condition. If the horse has a good burst of speed at the finish, you can let someone else set the pace, but do not let any horse get so far ahead that you will be unable to catch him. If, on the other hand, yours is a "one-pace" horse, you must keep him at that pace throughout the race regardless of your position at any time. If other horses drop back, continue to maintain the pace you have decided upon. A satisfactory race can be ridden just as well in front of the field as among or behind it, but a complete knowledge of the fastest pace your horse can maintain over the whole distance is essential.

Take the shortest way without interfering with other horses. Rounding a bend, begin your turn before the corner actually begins. Notice the other horses. If a certain horse jumps to the left, see to it that you are on his right. Avoid being abreast of and on the outside of a hard puller going around the bend; he may run out and take you with him. At least you will prevent him from running out which he might otherwise have done.

If you are riding abreast of a tired horse or a careless jumper, force the pace at the finish. He must then lose his place or jump faster than he wishes to. Do not force your horse if he feels tired. He may not have his second wind as yet. Wait for him to get it and then "make your bid." If you force him at this crucial moment, you may get a fall. Do not fight continually with a hard pulling horse. Let him go his own pace for a bit, then ease him up to the pace you want. He will go all the better for this.

The best time to take hold of your horse is directly after jumping a fence. *If your horse blunders and loses ground, do not at once hustle him into the place he has lost, as this will unsettle him.* Let him make up his distance gradually, when he has recovered.

Always try to get a good start. It is so much ground gained which the others must make up, and *in no way alters your waiting tactics should you have decided upon them.* If you have been fouled and wish to raise an objection, keep it to yourself until you get back to the waiting room and can place it before the proper authority. This will give you time to calm down and enable you to explain yourself clearly.

Check your equipment before the race and do not be late at the post or the scales. If you cannot be in the first four, you might as well be the last, so do not force a tired horse into a position which means nothing. He will be all the better for being allowed to finish quietly. Weigh out with your saddle complete and without your whip. If you happen to be underweight when weighing in, remember you can claim the weight of your bridle. Although short of losing a stirrup or other piece of equipment, this question should never arise. It is almost needless to say that you should never dismount after the race if you are in the first four without a direct order from an official.

Avoid sharp spurs. They should be unnecessary. If you can avoid using your whip, do so. In any case, the last fifty yards or so, or just before a finish, are the only cases when it can be of any service, and then only if properly used. To use your whip and sit still at the same time is extremely difficult. Those who can do it require no advice on the subject of riding races as it comes only from long practice. The beginner will usually do better to ride his horse home with his hands.

The Rider is in Balance and Does Not Interfere With His Horse.

An Excellent Position of the Rider Over the Top of a Jump.

An Illustration of the Correct Position of the Rider as the Horse Contacts the Ground in Landing Over an Obstacle.

XENOPHON PRESS LIBRARY

www.XenophonPress.com
Xenophon Press is dedicated to the preservation
of classical equestrian literature.
We bring both new and old works to English-speaking riders.

30 Years with Master Nuno Oliveira, Henriquet 2011
A Journey Through the Horse's Body, Fritz 2021
A Rider's Survival from Tyranny, de Kunffy 2012
Another Horsemanship, Racinet 1994
Austrian Art of Riding, Poscharnigg 2015
Broken or Beautiful: The Struggle of Modern Dressage, Barbier/Conrod 2021
Classic Show Jumping: the de Nemethy Method, de Nemethy 2016
Classical Dressage With Anja Beran by Anja Beran, Beran 2021
Divide and Conquer Book 1, Lemaire de Ruffieu 2016
Divide and Conquer Book 2, Lemaire de Ruffieu 2017
Dressage for the 21st Century, Belasik 2001
Dressage in the French Tradition, Diogo de Bragança 2011
Dressage Principles and Techniques: A Blueprint for the Serious Rider, Tavora 2018
Dressage Principles Illuminated, Expanded Edition, de Kunffy 2021
Dressage Sabbatical: A Year of Riding with Classical Master Paul Belasik, Caslar 2016
École de Cavalerie Part II, Robichon de la Guérinière 1992, 2015
Elements of Dressage, von Ziegner 2021
Equestrian Art Collected Works, Nuno Oliveira 2021
Equine Osteopathy: What the Horses Have Told Me, Giniaux 2014
Equitation, Bussigny 2021
Fragments from the Writings of Max Ritter von Weyrother, Fane 2017
François Baucher: The Man and His Method, Baucher/Nelson 2013
General Chamberlin: America's Equestrian Genius, Matha 2020
Great Horsewomen of the 19th Century in the Circus, Nelson 2015
Gymnastic Exercises for Horses Volume II, Eleanor Russell 2013
H. Dv. 12 German Cavalry Manual of Horsemanship, Reinhold 2014
Handbook of Jumping Essentials, Lemaire de Ruffieu 2015
Handbook of Riding Essentials, Lemaire de Ruffieu 2015
Healing Hands, Giniaux, DVM 1998

Horse Training: Outdoors and High School, Beudant 2014
I, Siglavy, Asay 2018
Horsemanship & Horsemastership Volume 1, US Cavalry 2021
Horsemanship Training Films 3 DVD set, US Cavalry 2021
Learning to Ride, Santini 2016
Legacy of Master Nuno Oliveira, Millham 2013
Lessons in Lightness: Expanded Edition, Mark Russell 2019
Methodical Dressage of the Riding Horse, Faverot de Kerbrech 2010
Military Equitation or, A Method of Breaking Horses, and Teaching Soldiers to Ride, Pembroke, and *A Treatise on Military Equitation*, Tyndale 2018
My Horses have Something to Say, de Wispelaere 2021
Principles of Dressage and Equitation, a.k.a. Breaking and Riding, Fillis 2017
Racinet Explains Baucher, Racinet 1997
Releasing the Jaw, Poll, and Neck DVD, Mark Russell 2021
Riding and Schooling Horses, Chamberlin 2020
Riding by Torchlight, Cord 2019
Riding in Rhyme, Davies 2021
Schooling Exercises in Hand, Hilberger 2021
Science and Art of Riding in Lightness, Stodulka 2015
The Art of Riding a Horse, D'Eisenberg 2015
The Art of Traditional Dressage, Volume I DVD, de Kunffy 2013
The Chamberlin Reader, Chamberlin/Matha, 2020
The de Nemethy Method: A training seminar, 8 DVD set, de Nemethy 2019
The Ethics and Passions of Dressage Expanded Edition, de Kunffy 2013
The Forward Impulse, Santini 2016
The Gymnasium of the Horse, Steinbrecht 2018
The Horses, a novel, Walker 2015
The Italian Tradition of Equestrian Art, Tomassini 2014
The Maneige Royal, de Pluvinel 2010, 2015
The New Method of Dressing Horses, Cavendish 2020
The Portuguese School of Equestrian Art, de Oliveira/da Costa 2012
The Spanish Riding School & Piaffe and Passage, Decarpentry 2013
The Spanish Riding School: The Miracle of the White Horse DVD, US Lipizzan Association 2021
To Amaze the People with Pleasure and Delight, Walker 2015
Total Horsemanship, Racinet 1999
Training Hunters, Jumpers, and Hacks, Chamberlin 2019
Training Your Foal, Ettl 2021
Training with Master Nuno Oliveira, 2 DVD set, Eleanor Russell 2016
Truth in the Teaching of Master Nuno Oliveira, Eleanor Russell 2015
Wisdom of Master Nuno Oliveira, de Coux 2012

www.ingramcontent.com/pod-product-compliance
Lightning Source LLC
Chambersburg PA
CBHW082104280426
43661CB00089B/858